ED-2525

TK
2000
.N368
1984

Electric Machines and Transformers

SYED A. NASAR

Department of Electrical Engineering
University of Kentucky

ELECTRIC MACHINES AND TRANSFORMERS

Macmillan Publishing Company
New York

Collier Macmillan Publishers
London

Copyright © 1984, Macmillan Publishing Company, a
division of Macmillan, Inc.

Printed in the United States of America

All rights reserved. No part of this book may be
reproduced or transmitted in any form or by any means,
electronic or mechanical, including photocopying,
recording, or any information storage and retrieval
system, without permission in writing from the Publisher.

Macmillan Publishing Company
866 Third Avenue, New York, New York 10022

Collier Macmillan Canada, Inc.

Library of Congress Cataloging in Publication Data

Nasar, S. A.
   Electric machines and transformers.
   Includes index.
   1. Electric machinery. 2. Electric transformers.
I. Title.
TK2000.N368   1984   621.31'042   83-9416
ISBN 0-02-385950-4

Printing: 1 2 3 4 5 6 7 8   Year: 4 5 6 7 8 9 0 1 2

ISBN   0-02-385950-4

Dedicated to my Daughter
Sajida

PREFACE

This book is intended for a first course in electric machines and transformers. The level of advanced mathematics has been kept to a minimum, and the emphasis is on phenomena and physical characteristics of machines. Their quantitative evaluations are made only in terms of simple mathematics, not beyond the high school level. However, concepts such as differentiation and integration are used rather loosely and may be skipped by the uninitiated without a loss of continuity. Throughout the text, numerous worked out examples have been used to illustrate the theories developed. In all instances, fundamentals are emphasized and practical examples and data are presented wherever possible.

It seems logical to start our discussions with magnetic circuits—an integral component of all electric machines. Thus, magnetic materials, their properties and circuit calculations are given in Chapter 1. Transformers—in particular power transformers—are discussed in Chapter 2. A number of photographs are included to illustrate the size, appearance, and components of certain transformers. Chapters 3, 4, and 5 respectively deal with dc machines, synchronous machines, and polyphase induction motors. A separate chapter—Chapter 6—is devoted to small (fractional horsepower) ac motors. Following the discussion of various motors in Chapters 3 through 6, electric drives and control of electric motors are presented in Chapter 7. Because solid-state controllers are now in common use in the industry, such controllers are presented in some detail.

Finally, the book ends with "Practical Considerations in Electric Machines" in Chapter 8.

For prerequisites, the reader is expected to be familiar with the steady-state analyses of dc and ac circuits.

Acknowledgment is made to John Wiley and Sons, Inc. for granting permission to use certain material from *Electric Machines and Electromechanics* by S. A. Nasar and L. E. Unnewehr, Wiley, 1979.

S. A. N.

CONTENTS

1 MAGNETIC CIRCUITS — 1

- 1.1 Magnetic Field and Flux — 1
- 1.2 Sources of Magnetic Flux: The Magnetomotive Force — 5
- 1.3 Magnetic Materials and Permeability — 9
- 1.4 Hysteresis: An Experiment with a Ferromagnetic Material — 14
- 1.5 DC Operation of Magnetic Circuits — 15
- 1.6 Differences Between Magnetic and Electric Circuits — 23
- 1.7 AC Operation of Magnetic Circuits — 24
- 1.8 Reduction of Iron Losses — 27
- 1.9 Inductance — 32
- 1.10 Magnetic Coupling of Electric Circuits: Mutual Inductance — 35
- 1.11 Energy Stored in Magnetic Circuits — 37
- 1.12 Equivalent Circuit of a Core Excited by an AC mmf — 39
- Problems — 41

2 TRANSFORMERS — 47

- 2.1 Introduction — 47

2.2	Transformer Construction	48
2.3	Transformer Classification	52
2.4	Principle of Operation: The Ideal Transformer	53
2.5	Voltage, Current, and Impedance Transformations	57
2.6	The Nonideal Transformer	61
2.7	Performance Characteristics of a Transformer	65
2.8	Approximate Equivalent Circuits	68
2.9	Equivalent Circuits from Test Data	71
2.10	Transformer Polarity	75
2.11	Transformers in Parallel	77
2.12	Three-Phase Transformer Connections	80
2.13	Certain Special Transformer Connections	85
2.14	Parallel Operation of Three-Phase Transformers	87
2.15	Autotransformers	88
2.16	Three-Winding Transformers	92
2.17	Instrument Transformers	93
2.18	Third Harmonics in Transformers	95
	Problems	98

3 DC MACHINES 103

3.1	Introduction	103
3.2	The Faraday Disk and Faraday's Law	104
3.3	The Heteropolar or Conventional DC Machine	107
3.4	Constructional Details	111
3.5	Classification According to Forms of Excitation	119
3.6	Performance Equations	121
3.7	Armature Reaction	126
3.8	Reactance Voltage and Commutation	129
3.9	Building-up of Voltage in a Shunt Generator	131
3.10	Generator Characteristics	133
3.11	Motor Characteristics	136
3.12	Starting and Control of Motors	137
3.13	Losses and Efficiency	142
3.14	Tests on DC Machines	144
3.15	Certain Applications	147
3.16	Parallel Operation of DC Generators	147
	Problems	151

4 SYNCHRONOUS MACHINES — 155

- 4.1 Introduction — 155
- 4.2 Some Construction Details — 156
- 4.3 Magnetomotive Forces and Fluxes Due to Armature and Field Windings — 163
- 4.4 Synchronous Speed — 167
- 4.5 Synchronous Generator Operation — 171
- 4.6 Performance of a Round-Rotor Synchronous Generator — 173
- 4.7 Synchronous Motor Operation — 176
- 4.8 Performance of a Round Rotor Synchronous Motor — 178
- 4.9 Salient-Pole Synchronous Machines — 181
- 4.10 Parallel Operation — 184
- 4.11 Determination of Machine Constants — 190
- Problems — 195

5 INDUCTION MOTORS — 199

- 5.1 Introduction — 199
- 5.2 Operation of a Three-Phase Induction Motor — 202
- 5.3 Slip — 203
- 5.4 Development of Equivalent Circuits — 204
- 5.5 Performance Calculations — 207
- 5.6 Approximate Equivalent Circuit From Test Data — 212
- 5.7 Performance Criteria of Induction Motors — 215
- 5.8 Speed Control of Induction Motors — 217
- 5.9 Starting of Induction Motors — 220
- 5.10 Induction Generators — 225
- 5.11 Energy-Efficient Induction Motors — 228
- Problems — 231

6 SMALL AC MOTORS — 235

- 6.1 Introduction — 235
- 6.2 Single-Phase Induction Motors — 236
- 6.3 Small Synchronous Motors — 244
- 6.4 AC Commutator Motors — 249
- 6.5 Two-Phase Motors — 253
- 6.6 Stepper Motors — 255
- Problems — 260

7 ELECTRIC DRIVES 263

7.1	General Remarks	263
7.2	Power Semiconductors	265
7.3	Controllers for DC Motors	270
7.4	Controllers for AC Motors	283
7.5	Choice of Motor-Drive System	293
	Problems	295

8 PRACTICAL CONSIDERATIONS IN ELECTRIC MACHINES 299

8.1	General Remarks	299
8.2	Nameplate Ratings	299
8.3	Efficiency and Duty Cycle	302
8.4	Ratings and NEMA Classifications of Electric Machines	306
8.5	Economic Considerations	308
8.6	Thermal Considerations	310
8.7	Motor Selection	313
8.8	Varying Load Applications	315
8.9	Insulation Considerations	317
	Problems	320

REFERENCES		323
APPENDIX A	Wire Table	325
APPENDIX B	Conversion Factors	327
APPENDIX C	Data on DC Machines	329
APPENDIX D	Data on AC Machines	333
APPENDIX E	List of Symbols	337
APPENDIX F	Answers to Selected Problems	345
INDEX		347

Electric Machines and Transformers

CHAPTER 1

Magnetic Circuits

1.1
MAGNETIC FIELD AND FLUX

Electric machines and transformers are made up of interlinked electric and magnetic circuits. Electric currents flow through electric circuits and magnetic fluxes (which we will define in this chapter) flow through magnetic circuits. Interactions between electric currents and magnetic fluxes result in the processes of electromechanical energy conversion, as in electric motors and generators, and electric energy transfer as in transformers. Thus we see that magnetic circuits play an important role in the operation of electric machines and transformers. In this chapter we review the concepts pertinent to magnetic circuits and present methods of analyzing them.

We are familiar with the phenomenon of magnetism through the compass, which indicates the direction of the earth's magnetic poles and thus helps us in navigation. We also encounter magnets in toys and in household fixtures. Such magnets are permanent magnets. These are pieces of iron or certain metal alloys, or ceramic-like porcelain (known as ceramic magnets or ferrite magnets), and have the property of attracting or repelling each other. Alnico, Indox, Cunife, and Hicorex are some of the trade names of commercially available permanent magnets. Historically, we have been aware of magnets for centuries.

Although we will discuss permanent magnets later in this chapter, let us consider a common and very old experiment pertaining to magnets. If a bar magnet is placed under a sheet of paper and iron filings are sprinkled over it, the filings cling to the magnet and form a definite pattern around it.

A typical pattern is shown in Fig. 1.1. The filings act like tiny magnets attracting each other. These are most dense near the two ends, or *poles,* of the magnet. Magnetic poles always occur in pairs of opposite kind: north and south. Magnetic monopoles have not been found to exist. It has been observed experimentally that like poles repel and opposite poles attract one another. Even if two magnets are not in contact, a force acts between them. Conceptually, this phenomenon is action at a distance, and can be explained by postulating that every magnet is surrounded by a *magnetic field.* A magnet placed in a magnetic field is subject to a force produced by the field. An alternative concept of a magnetic field is that if a moving charge in a region experiences a force, by virtue of its motion in a region of space, then a magnetic field exists in the region. Because the motion of electrons (negative electrical charges) constitutes electric current, a current-carrying conductor located in a certain way in a magnetic field will experience a force, as discussed below.

A magnetic field may be represented by field lines. Roughly speaking, the iron-filing pattern of Fig. 1.1 corresponds to the field lines generated by the bar magnet. These field lines are also known as *magnetic lines of force.* The lines of force can be traced with the aid of a compass needle. The lines are continuous, begin at one pole, go to the other pole, and return to the point of origin through

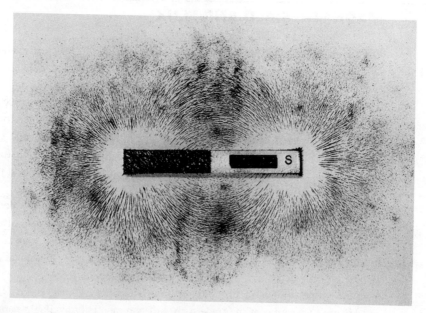

FIGURE 1-1. Iron filings pattern around a bar magnet.

the body of the magnet. These lines may be considered to correspond to magnetic flux lines. The magnet flux lines are measured in *webers* (Wb). However, we need a better and precise definition of magnetic flux. We recall from the preceding paragraph that a current-carrying conductor experiences a force in a magnetic field (regardless of how the magnetic field has been produced). This fact is enunciated as *Ampère's* (force) law, according to which a current-carrying conductor, of length l meters (m) and carrying a current I amperes (A), located in a magnetic field such that the conductor and the magnetic lines of force are mutually perpendicular, will experience a force, F, as given by

$$F = BlI \quad \text{newtons (N)} \tag{1.1}$$

In (1.1) B is known as the *magnetic flux density* and has the direction of magnetic lines of force. The SI unit of B is the tesla (T). The direction of F, as given by (1.1), is perpendicular to the B and lI.* Specifically, the *left-hand rule* as illustrated in Fig. 1.2 gives the mutual relationships between the directions of F, B, and lI.

Let the magnetic flux density, B, be uniform (or constant) in a certain region and be perpendicular to a surface area A meter² (m²), then the *magnetic flux*, φ, is defined by

$$\varphi = BA \quad \text{webers (Wb)} \tag{1.2}$$

If the surface area makes an angle θ with the B, the magnetic flux, or simply flux, is given by

$$\varphi = BA \sin \theta \tag{1.3}$$

This relationship is depicted in Fig. 1.3.

*We use lI to indicate that the orientation of the conductor, of length l, determinates the orientation of the current I.

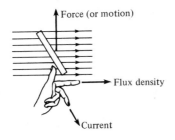

FIGURE 1-2. The left-hand rule.

1.1 MAGNETIC FIELD AND FLUX

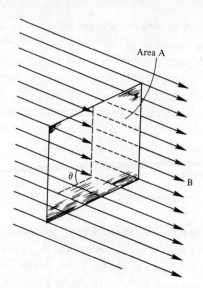

FIGURE 1-3. B and A at an angle Θ.

EXAMPLE 1.1

A 2-m-long straight conductor, located at right angles to a uniform magnetic field (that is, the flux density is constant throughout the region), while carrying a 5-A current experiences a force of 7 N. What is the flux density in the region?

SOLUTION

This problem involves a direct application of (1.1), from which we have

$$7 = B \times 2 \times 5$$

Hence

$$B = 0.7 \text{ T}$$

∎

EXAMPLE 1.2

A circular loop of 2 m diameter is located in a magnetic field of 0.5 T. The plane of the loop makes a 60° angle with the flux lines. Determine the flux passing through the loop.

SOLUTION

In this problem we will use (1.3). Thus the area of the loop,

$$A = \frac{\pi}{4} \times 2^2 = \pi \text{ m}^2$$

and

$$\theta = 60°$$

Hence from (1.3),

$$\varphi = 0.5 \times \pi \times \sin 60° = 1.36 \text{ Wb}$$

1.2 SOURCES OF MAGNETIC FLUX: THE MAGNETOMOTIVE FORCE

From Section 1.1 it is clear that magnets are sources of magnetic flux. One form of a magnet is the permanent magnet, such as the one used in a compass, as mentioned earlier. The other form of a magnet is the electromagnet. In contrast to a permanent magnet, an electromagnet always requires for its operation an electric current, which must flow in an appropriate electric circuit. As we will see later, such an electric circuit is often formed in the shape of a coil. Magnetic fluxes produced by a cylindrical rod-shaped permanent magnet and by a coil wound on a cardboard tube are shown in Fig. 1.4(a) and (b), respectively. The flux lines in a permanent magnet are assumed to emerge from the north (N) pole and enter back into the magnet at the south (S) pole. This convention determines the directions of the flux lines, as shown in Fig. 1.4(a). Similarly, the directions of flux lines produced by the coil of Fig. 1.4(b) are also illustrated in Fig. 1.4(b). These are determined by the *right-handed screw rule*; that is, if we turn a right-handed screw such that it traverses in the direction of the current, the magnetic flux lines are in the direction of the rotation of the screw, as shown in Fig. 1.5(a). Figure 1.5(b) and (c) show two other configurations of current-carrying conductors and the resulting fluxes. Notice from Fig. 1.5(b) that concentric flux lines produced by a current in a straight conductor are drawn in accordance with the right-hand rule—if the thumb points to the direction of the current, the four fingers grabbing the conductor point to the direction of the flux lines. Applying the rules illustrated in Fig. 1.5(a) and (b) yields the direction of the flux lines due to a loop of current.

The magnetic flux produced by a current-carrying conductor is evaluated in terms of its *magnetomotive force,* or *mmf*. For an N-turn coil, such as the one of Fig. 1.6 carrying a current I, the mmf, $\mathcal{F}$, is defined by

$$\mathcal{F} = NI \quad \text{At} \tag{1.4}$$

where At stands for ampere-turns, the unit of mmf. Notice from Fig. 1.6 that

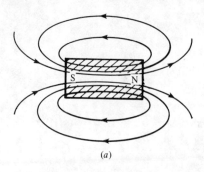

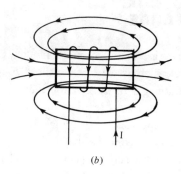

FIGURE 1-4. (a) flux lines from a cylindrical permanent magnet; (b) flux lines produced by an electromagnet.

the flux lines have a mean length l. We then say that the mmf is dropped over the length l, and we define the *magnetic field intensity*, H, as the mmf per unit length. Explicitly, from (1.4) we have

$$H = \frac{NI}{l} \quad \text{At/m} \tag{1.5}$$

In Figs. 1.4 and 1.6 we have shown that magnetic fluxes follow certain paths. A path for the flow of magnetic flux constitutes a *magnetic circuit*, just as an electric circuit provides a path for the flow of electric current. There are similarities and differences between magnetic and electric circuits. Based on the similarities, we will develop analytical methods for the study of magnetic circuits, keeping in mind that we must account for the differences as well. One of the major differences between magnetic and electric circuits is that the material properties of most magnetic circuits are significantly different from the material properties of electric circuits. Let us now pursue this difference in some detail.

EXAMPLE 1.3

For the magnetic circuit of Fig. 1.6, we have the following data: $N = 1200$ turns, $l = 20$ cm, and $A = 8$ cm² (which is the cross-sectional area of the

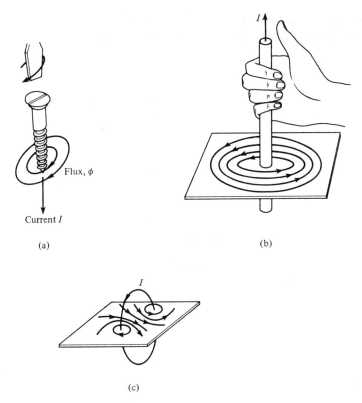

FIGURE 1-5. (a) right-hand screw rule for current-flux relationship; (b) flux lines due to a straight conductor; (c) flux lines due to a loop of current.

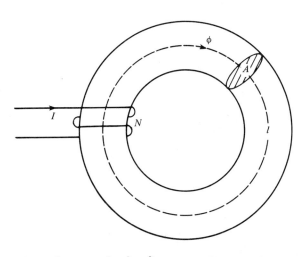

FIGURE 1-6. A magnetic circuit.

1.2 SOURCES OF MAGNETIC FLUX

ring). The current in the ring is gradually increased and the flux thus produced is measured experimentally, yielding the following data:

I (A)	50	100	150	200
φ (mWb)	0.3016	0.6032	0.9048	1.206

Obtain a relationship between B and H (a) graphically and (b) analytically.

SOLUTION

Using (1.5) and the table above, we convert I into H. From (1.2) and the table above, we change φ into B. Hence we obtain:

(a)

$H = NI/l$ (At/m)	300,000	600,000	900,000	1,200,000
$B = \varphi/A$ (T)	0.377	0.754	1.131	1.508

The results are plotted in Fig. 1.7.

(b) Let

$$B = kH$$

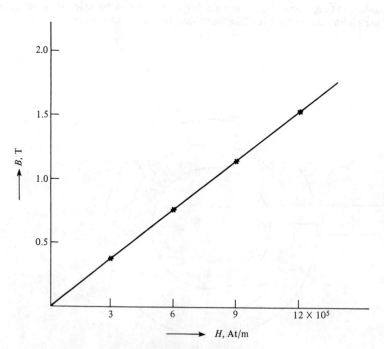

FIGURE 1-7. BH-curve for material of Example 1.3.

where k is a constant. Then from the data above we have

$$0.377 = 300{,}000k$$

or

$$k = 1.2567 \times 10^{-6}$$

Hence

$$B = 1.2567 \times 10^{-6} H \quad \text{T}$$

which may also be written as

$$B = 4\pi \times 10^{-7} H \quad \text{T}$$

The quantity $4\pi \times 10^{-7}$ is denoted by μ_0 in the study of magnetic circuits, as we shall presently see. ∎

1.3 MAGNETIC MATERIALS AND PERMEABILITY

We recall from the discussions of Sections 1.1 and 1.2 that the magnetic flux density, B, is dependent on the magnet field intensity, H. In the particular case of Example 1.3 we found this dependence to be linear. In general, the variation of B with H may not be linear, and we have

$$B = \mu H \tag{1.6}$$

where μ may itself depend on H. The quantity μ is known as the *permeability* of the medium in which B and H exist. The SI units of permeability are henries per meter (H/m). For air (or free space) the permeability is a constant μ_0 which is given by

$$\mu_0 = 4\pi \times 10^{-7} \text{H/m} \tag{1.7}$$

Notice from (1.7) that μ_0 has the same value as obtained in Example 1.3, indicating that the magnetic circuit of Fig. 1.6 consists of air only. We used rather realistic numbers for the data of Example 1.3. Numbers such as 100 A current, or 120,000 At mmf, indicate relatively large quantities. We needed such a large mmf to produce a reasonably small flux density of about 0.75 T in air. However, (1.6) indicates that larger flux densities can be produced for smaller mmf's (or magnetic field intensities) if we use materials having larger values of permeability. Iron is one such material. Its large permeability is the primary reason for its use in the magnetic circuits of electric machines and transformers.

For use in magnetic circuits, we may divide materials constituting these circuits into two groups. In one group, we may include those materials whose permeability approximately equals that of free space, μ_0. These are generally nonmagnetic materials. Examples of such materials are air, copper, and wood. To this group also belong the materials called *paramagnetic*—having a permeability slightly greater than μ_0—and *diamagnetic*, whose permeability is slightly less than μ_0. The second group includes those materials whose permeability is considerably greater than μ_0. This group mainly includes *ferromagnetic* and *ferrimagnetic* materials. Ferromagnetic materials are further subdivided into *hard* and *soft* materials. This classification roughly corresponds to the physical hardness of the material. Soft ferromagnetic materials include iron and most soft steels. Hard ferromagnetic materials include the permanent magnet materials such as alnicos and samarium–cobalt alloys. Ferrimagnetic materials are classified as ferrites and have much larger resistivities than those of ferromagnetic materials.

There are two significant characteristics of ferromagnetic materials such as iron with regard to their applications in magnetic circuits. First, the permeability of iron is considerably greater than the permeability of free space. Often, we express this permeability relationship in terms of *relative permeability*, μ_r:

$$\mu = \mu_r \mu_0 \quad \text{H/m} \tag{1.8}$$

where μ is the permeability of the material (iron, in this case) and $\mu_0 = 4\pi \times 10^{-7}$ H/m, the permeability of air. Second, the relationship of B to H (for iron) is nonlinear; that is, it cannot be represented by a straight line as in Fig. 1.7 (for air). A typical BH relationship for iron used in some magnetic circuits is shown in Fig. 1.8.

Let us examine Fig. 1.8 in some detail. We observe that no single value of permeability can be used to characterize the material. In such a case, it is convenient to divide the BH curve into three operating regions, labeled I, II, and III in Fig. 1.8. Although very few magnetic circuits are likely to operate in region I, we define the permeability for this region as the *initial permeability*, μ_i, which is given by

$$\mu_i = \lim_{H \to 0} \frac{B}{H} \tag{1.9}$$

where "$\lim_{H \to 0}$" reads "the limit as H tends to zero," implying that we are considering region I. For regions II and III, the permeability is simply the ratio

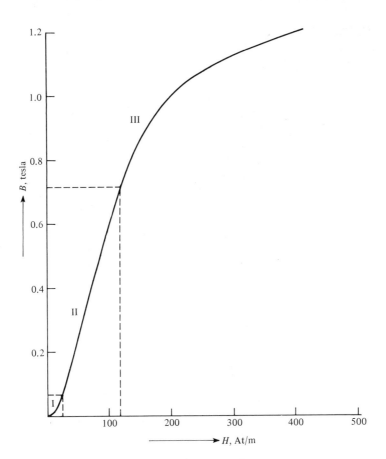

FIGURE 1-8. The BH-curve from an iron sample.

B/H. However, sometimes region II can be approximated by a straight line, in which case the permeability is constant for region II. In region III, a large change in H results in a relatively small change in B, and the material (iron) approaches *saturation*, where the magnetic flux density cannot increase beyond a certain level. Consequently, the BH curve is also known as the *saturation curve* or the *magnetization curve*. Region III is essentially the nonlinear operating region. In this region the permeability is sometimes considered as the *incremental* or *differential* permeability, μ_d, which is defined by

$$\mu_d = \frac{\Delta B}{\Delta H} \qquad (1.10)$$

where ΔB and ΔH are small changes in B and H, respectively.

We now illustrate the permeability calculations by the following examples.

1.3 MAGNETIC MATERIALS AND PERMEABILITY

EXAMPLE 1.4

The BH curve for an iron rod is shown in Fig. 1.8. (a) Determine the permeability for the region II, and (b) evaluate the incremental permeability if H varies from 160 to 200 At/m.

SOLUTION

(a) For region II the BH curve is practically a straight line. On this straight line, one pair of BH values are:

$$H = 100 \text{ At/m}$$
$$B = 0.6 \text{T}$$

Hence

$$\mu = \frac{B}{H} = \frac{0.6}{100} = 6 \times 10^{-3} \text{ H/m}$$

(b) The given values of H lie in region III, from which we have

$$\text{at } H = 160 \text{ At/m}$$
$$B = 0.84 \text{ T}$$
$$\text{at } H = 200 \text{ At/m}$$
$$B = 1.0 \text{ T}$$

Therefore,

$$\mu_d = \frac{\Delta B}{\Delta H} = \frac{1 - 0.84}{200 - 160} = 4 \times 10^{-3} \text{ H/m}$$

■

EXAMPLE 1.5

Find the relative permeability of iron operating in (a) region II and (b) at $H = 200$ At/m of the BH curve of Fig. 1.8.

SOLUTION

(a) From Example 1.4 we have, for region II,

$$\mu = 6 \times 10^{-3} \text{ H/m}$$

Substituting this value and $\mu_0 = 4\pi \times 10^{-7}$ in (1.8), we obtain

$$\mu_r = \frac{6 \times 10^{-3}}{4\pi \times 10^{-7}} = 4775$$

(b) From Fig. 1.8 we have

$$B = 1.0 \text{ T}$$
$$\text{at} \quad H = 200 \text{ At/m}$$

Therefore,

$$\mu = \frac{1}{200}$$

and

$$\mu_r = \frac{1}{200 \times 4\pi \times 10^{-7}} = 3979$$

Figure 1.9 shows the *BH* characteristics of sheet steel, cast steel, and cast iron, which are three of the most common materials used in magnetic circuits of electric machines. We will use these curves in making magnetic circuit calculations later in this chapter.

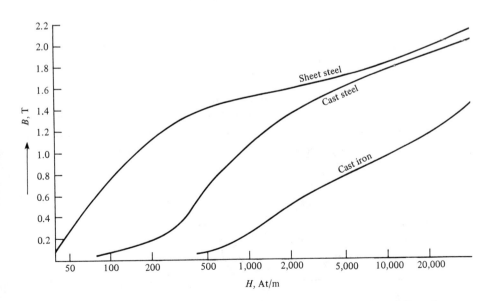

FIGURE 1-9. BH-Characteristics of Sheet Steel, Cast Steel, and Cast Iron.

1.3 MAGNETIC MATERIALS AND PERMEABILITY

1.4 HYSTERESIS: AN EXPERIMENT WITH A FERROMAGNETIC MATERIAL

In Fig. 1.10 we show a magnetic circuit made of a ferromagnetic material. On the magnetic circuit we have wound two coils, with which we make the connections as shown. We assume that the magnetic material is virgin in the sense that it has never been exposed to a magnetic field. The aim of the experiment is to obtain the BH characteristic of the material. With the switch open, the current in the N-turn coil is zero and so is the magnetic flux. Thus we start at point 0 in Fig. 1.11. Next, we close the switch at position 1 and gradually increase the current in the coil by adjusting the resistance R. For every value of the current we record the flux, as measured by a flux-measuring instrument. We obtain the curve $0a$, and at point a, the core reaches saturation. At this point we begin to decrease the current while measuring the flux for every value of the current. We thus trace the curve aB_r. Notice from Fig. 1.11 that in decreasing the current we did not trace back $a0$. The value of the flux density B_r is the *remanent flux density*; that is, it indicates the degree of magnetization of the core material. From Fig. 1.11 it is clear that at B_r, H (or I) is zero. Now we reverse the current in the coil by putting the switch in position 2. As we increase the current in the negative direction, the flux density decreases and reaches zero at a certain value of $-I$, with corresponding field intensity H_c. This field intensity is known as the coercive field intensity. The corresponding mmf is the *coercive force*. The significance is that we need an mmf equal to the coercive force to demagnetize the core, or reduce the residual (or remanent) magnetism to zero.

Next, if we continue to increase the current in the negative direction, we trace the curve $H_c b$ and reach the negative saturation flux density $-B_s$. The negative sign implies that the flux lines are now opposite in direction to $+B_s$. Finally, we decrease the current to trace bc where the current is zero. We now put the switch back to position 1, increase the current in the positive direction (while measuring the flux), and trace the portion of the curve marked $bcda$. The loop $abcda$ is known as the *hysteresis loop*. The phenomenon of hysteresis is a characteristic of magnetic materials used in electric machines.

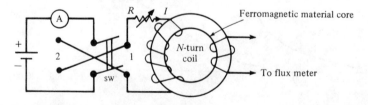

FIGURE 1-10. Experimental arrangement for obtaining the BH-characteristic.

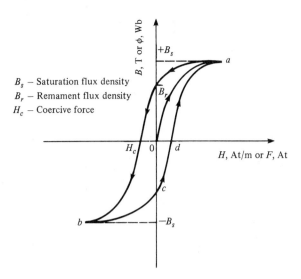

FIGURE 1-11. Hysteresis loop and certain pertinent quantities.

From Fig. 1.11 it is clear that hysteresis is present only when the magnetic material is exposed to a time-varying magnetic field. A common example of a source of a time-varying magnetic field is the sinusoidal alternating-current (ac) mmf, that is, a coil carrying ac. Thus if the mmf acting on a magnetic material is ac, the *BH* curve takes the form of the symmetrical hysteresis loop of Fig. 1.11. The area within the loop is proportional to the energy loss (as heat) in the material per cycle. This energy loss is known as *hysteresis loss*. Therefore, in evaluating the losses in an electromagnetic device exposed to magnetic fields undergoing time variations, we must account for the hysteresis loss. In Section 1.6 we will return to this subject.

Table 1.1 shows the characteristics of a number of soft magnetic materials. The important data included are B_s, H_c, μ, and electrical resistivity. The Curie temperature is the point at which a material loses its magnetic properties.

1.5

DC OPERATION OF MAGNETIC CIRCUITS

In the preceding sections we have alluded to the similarities between magnetic and electric circuits. We now use these similarities in formulating some of the laws governing magnetic circuits. These laws, in turn, will aid us in analyzing magnetic circuits.

We refer to Fig. 1.12(a), which shows a magnetic circuit. Notice that the path of the flux, φ, is defined by the dashed line. We term the magnetic material of the magnetic circuit as the *core*. On the core, we have an *N*-turn coil, carrying

TABLE 1.1 Characteristics of Certain Soft Magnetic Materials

Trade Name	Principal Alloys	Saturation Flux Density (T)	H at B_{sat} (A/m)	Amplitude Permeability Max μ_m	Coercive Force, H_c (A/m)	Electrical Resistivity ($\mu\Omega$-cm)	Curie Temperature (°C)
48 NI	48% Ni	1.25	80	200,000		65	
Monimax	48% Ni	1.35	6,360	100,000	4.0	65	398
High Perm 49	49% Ni	1.1	80			48	
Satmumetal	Ni, Cu	1.5	32	240,000		45	398
Permalloy (sheet)	Ni, Mo	0.8	400	100,000	1.6	55	454
Moly Permalloy (powder)	Ni, Mo	0.7	15,900	125			
Deltamax	50% Ni	1.4	25	200,000	8	45	499
M-19	Si	2.0	40,000	10,000	28	47	
Silectron	Si	1.95	8,000	20,000	40	50	732
Oriented-T	Si	1.6	175			47	
Oriented M-5	Si	2.0	11,900	30,000	26	48	746
Ingot Iron	None	2.15	55,000		80	10.7	
Spermendur	49% Co, V	2.4	15,900	80,000	8	26	932
Vanadium Permendur	49% Co, V	2.3	12,700	4,900	92	40	925
Hyperco 27	27% Co	2.36	70,000	2,800	198	19	
Flake Iron	Carbonal power	≈0.8	5,200	5–130		10^5–10^{15}	
Ferrotron (powder)	Mo, Ni	(Linear)	(Linear)	5–25		10^{16}	
Ferrite	Mg, Zn	0.39	1,115	3,400	13	10^7	135
Ferrite	Mn, Zn	0.453	1,590	10,000	6.3	3×10^7	190
Ferrite	Ni, Zn	0.22	2,000	160	318	10^9	500
Ferrite	Ni, Al	0.28	6,360	400	143		500
Ferrite	Mg, Mn	0.37	2,000	4,000	30	1.8×10^8	210

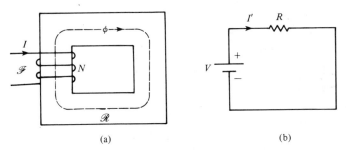

FIGURE 1-12. (a) a magnetic circuit; (b) DC electric circuit.

I', a direct current (dc), and resulting in an mmf $\mathscr{F} = NI$. This mmf is the source of the flux. In Fig. 1.12(b) we show a dc resistive circuit. The voltage V results in the current I'. The current, I', and voltage, V, are related to each other by Ohm's law:

$$V = RI' \qquad (1.11)$$

If we let V of the electric circuit correspond to $\mathscr{F}$ of the magnetic circuit, and I' to φ, then Ohm's law, (1.11), takes the form

$$\mathscr{F} = \mathscr{R}\varphi \qquad (1.12)$$

Comparing (1.11) and (1.12), we observe that $\mathscr{R}$ of the magnetic circuit is analogous to R, the resistance of the electric circuit. The quantity $\mathscr{R}$ is known as the *reluctance*. The reciprocal of reluctance is called *permeance*.

Based on the preceding analogy, we may now construct the relationships summarized in Table 1.2. In this table l is the length and A is the cross-sectional area of the path for the flow of current in the electric circuit or for the flux in the magnetic circuit. It may be verified from Table 1.2 that the unit of reluctance is H^{-1}.

Because φ is analogous to I and $\mathscr{R}$ is analogous to R, the laws for series- or parallel-connected resistors also hold for reluctances. Figure 1.13 shows four

TABLE 1.2 Analogy Between a Magnetic Circuit and a DC Electric Circuit

Magnetic Circuit	Electric Circuit
Flux, φ	Current, I
Mmf, $\mathscr{F}$	Voltage, V
Reluctance, $\mathscr{R} = l/\mu A$	Resistance, $R = l/\sigma A$
Permeance, $\mathscr{P} = 1/\mathscr{R}$	Conductance, $G = 1/R$
Permeability, μ	Conductivity, σ
Ohm's law, $\varphi = \mathscr{F}/\mathscr{R}$	Ohm's law, $I = V/R$

1.5 DC OPERATION OF MAGNETIC CIRCUITS

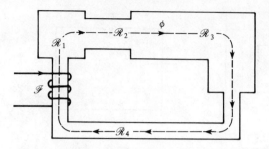

FIGURE 1-13. Reluctances in series.

reluctances connected in series. For such a magnetic circuit we have the following relationship:

Reluctances in series:

$$\mathcal{R}_{ser} = \mathcal{R}_1 + \mathcal{R}_2 + \mathcal{R}_3 + \cdots \qquad (1.13)$$

On the other hand, for reluctances in parallel, as in Fig. 1.14, we have the relationship:

Reluctances in parallel:

$$\frac{1}{\mathcal{R}_{par}} = \frac{1}{\mathcal{R}_1} + \frac{1}{\mathcal{R}_2} + \frac{1}{\mathcal{R}_3} + \cdots \qquad (1.14)$$

Let us now illustrate the applications of (1.13) and (1.14) by the following examples.

EXAMPLE 1.6

With reference to Fig. 1.13, show that (1.13) is a valid expression for reluctances in series.

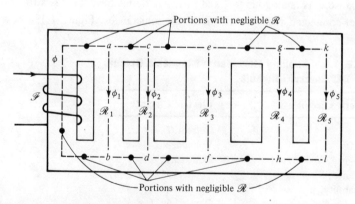

FIGURE 1-14. Reluctances in parallel.

SOLUTION

In Fig. 1.13 we notice that φ is the common flux through all the four reluctances. Hence the mmf drops across the reluctances are $\mathscr{F}_1 = \mathscr{R}_1\varphi$, $\mathscr{F}_2 = \mathscr{R}_2\varphi$, ... If the total mmf $\mathscr{F}$ is dropped across a net reluctance $\mathscr{R}_{ser}$, we have

$$\mathscr{F} = \mathscr{R}_{ser}\varphi = \mathscr{R}_1\varphi + \mathscr{R}_2\varphi + \mathscr{R}_3\varphi + \mathscr{R}_4\varphi$$

or

$$\mathscr{R}_{ser} = \mathscr{R}_1 + \mathscr{R}_2 + \mathscr{R}_3 + \mathscr{R}_4$$

which is the same as (1.13). ∎

EXAMPLE 1.7

Show that (1.14) is valid for reluctances in parallel.

SOLUTION

We refer to Fig. 1.14 and assume that reluctances are in parallel only and that the connecting branches joining these reluctances as shown in Fig. 1.14 do not offer any reluctance. Let $\mathscr{F}$ be the common mmf across $\mathscr{R}_1$, $\mathscr{R}_2$, ... Then the fluxes φ_1, φ_2, ... are given by

$$\varphi_1 = \frac{\mathscr{F}}{\mathscr{R}_1}, \quad \varphi_2 = \frac{\mathscr{F}}{\mathscr{R}_2}, \quad \cdots$$

If $\mathscr{R}_{par}$ is the total reluctance for the parallel combination and φ is the total flux, then

$$\frac{\mathscr{F}}{\mathscr{R}_{par}} = \varphi = \varphi_1 + \varphi_2 + \cdots = \frac{\mathscr{F}}{\mathscr{R}_1} + \frac{\mathscr{F}}{\mathscr{R}_2} + \cdots$$

Consequently,

$$\frac{1}{\mathscr{R}_{par}} = \frac{1}{\mathscr{R}_1} + \frac{1}{\mathscr{R}_2} + \cdots$$

which is the same as (1.14). ∎

EXAMPLE 1.8

In analyzing magnetic circuits such as the one shown in Fig. 1.15, we use the concept of *mean length*. For the circuit of Fig. 1.15(a), we designate the mean lengths by l's and the various areas of cross sections by A's. If $2A_1 = A_2 = 10$ cm², $l_1 = 3l_2 = 24$ cm and the relative permeability of the magnetic

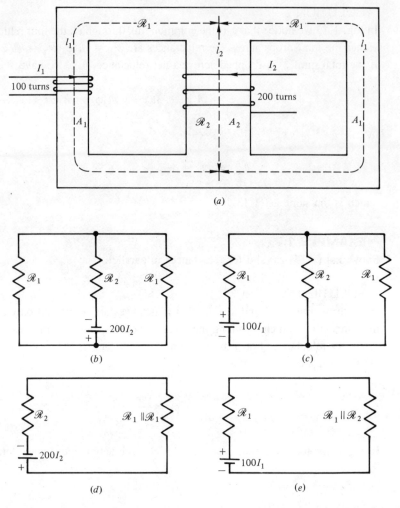

FIGURE 1-15. (a) A magnetic circuit; (b) electrical analog when 200-turn coil excites; (c) electrical analog when 100-turn coil excites; (d) further reduction of circuit (b); (e) further reduction of circuit (c).

material at a certain operating point is 500, calculate the net reluctance as seen by (a) the 200-turn coil and (b) the 100-turn coil. Also, show electrical analogs for the two cases. Both coils are not assumed to be excited at the same time. Neglect saturation.

SOLUTION

The electrical analogs are shown in Fig. 1.15(b) and (c), respectively, for the cases when only the 200-turn coil is excited and when only the 100-turn coil is excited. These analogs, in turn, are respectively reduced to those shown in Fig. 1.15(d) and (e). From the given data

$$\mu = \mu_r\mu_0 = 500 \times 4\pi \times 10^{-7} = 6.28 \times 10^{-4} \text{ H/m}$$

$$\mathcal{R}_1 = \frac{l_1}{\mu A_1} = \frac{24 \times 10^{-2}}{6.28 \times 10^{-4} \times \frac{10}{2} \times 10^{-4}} = 0.764 \times 10^{-6} \text{ H}^{-1}$$

$$\mathcal{R}_2 = \frac{l_2}{\mu A_2} = \frac{8 \times 10^{-2}}{6.28 \times 10^{-4} \times 10 \times 10^{-4}} = 0.127 \times 10^{-6} \text{ H}^{-1}$$

We now use these values of reluctance in Fig. 1.15(b)–(e).

(a) We have from Fig. 1.15(d) the reluctances as seen by the 200-turn coil as the sum: the two $\mathcal{R}_1$'s in parallel, the combination being in series with $\mathcal{R}_2$. Hence

$$\mathcal{R}_{(a)} = \tfrac{1}{2}\mathcal{R}_1 + \mathcal{R}_2 = 10^{-6}(\tfrac{1}{2} \times 0.764 + 0.127) = 0.509 \times 10^{-6} \text{ H}^{-1}$$

(b) From Fig. 1.15(e) we have the reluctance, as seen by the 100-turn coil, as the sum: $\mathcal{R}_1$ in parallel with $\mathcal{R}_2$, the combination being in series with $\mathcal{R}_1$. Hence

$$\mathcal{R}_{(b)} = \frac{\mathcal{R}_1\mathcal{R}_2}{\mathcal{R}_1 + \mathcal{R}_2} + \mathcal{R}_1 = \frac{10^{-12}(0.764 \times 0.127)}{10^{-6}(0.764 + 0.127)} + 10^{-6} \times 0.764$$

$$= 0.873 \times 10^{-6} \text{ H}^{-1} \qquad \blacksquare$$

EXAMPLE 1.9

A magnetic circuit containing an air gap is shown in Fig. 1.16. Such a magnetic circuit is known as a composite magnetic circuit. With the dimensions as marked, calculate the current in the coil to establish a 0.6-T flux density in the air gap. The iron core has the *BH* characteristic shown in Fig. 1.8.

SOLUTION

Let $\mathcal{F}$ be the mmf required to produce the 0.6-T flux density. This mmf overcomes the reluctance of the iron core and the air gap. We now determine these mmf's.

Iron core: At $B = 0.6$ T, from Fig. 1.8, we have $H = 100$ At/m and $l = 20$ cm (given). Thus

$$\mathcal{F}_{core} = Hl = 100 \times 0.2 = 20 \text{ At}$$

Air gap: Since

$$\mathcal{R} = \frac{g}{\mu_0 A}$$

1.5 DC OPERATION OF MAGNETIC CIRCUITS

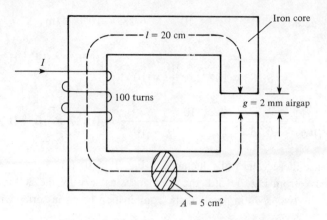

FIGURE 1-16. Magnetic circuit of Example 1.9.

and

$$\varphi = BA$$

we have

$$\mathcal{F}_{air} = \mathcal{R}\varphi = \frac{gBA}{\mu_0 A} = \frac{gB}{\mu_0}$$

$$= \frac{2 \times 10^{-3} \times 0.6}{4\pi \times 10^{-7}} = 955 \text{ At}$$

Hence

$$\mathcal{F} = \mathcal{F}_{core} + \mathcal{F}_{air} = 20 + 955 = 975 \text{ At} = NI$$

But $N = 100$; therefore,

$$I = \frac{975}{100} = 9.75 \text{ A}$$

Before we proceed to the next topic, let us review some of the results of Example 1.9. First, the mmf of the air gap is about 50 times the mmf of the iron core, even though the iron core mean length is 100 times that of the air gap. In other words, for a given length the mmf required for the ferromagnetic portion of a composite magnetic circuit is almost negligible compared to that for the air gap. Consequently, in most calculations, as a first approximation, we often assume that the permeability of iron is infinity. Recall also from Example 1.5 that $\mu_r = 4000$ for the iron sample, implying that $\mu_{iron} \gg \mu_0$. If we had entirely neglected the iron core in Example 1.9, the required current would have been 9.55 A (which is off by 0.2 A from the exact answer). Second, notice in our calculations for $\mathcal{F}_{core}$ in Example 1.9 that we did not calculate any reluctances since the BH or $\varphi\mathcal{F}$ relationship was available in the form of Fig. 1.8, from which $\mathcal{F}$ was directly obtained. Furthermore, computations of permeabilities in the nonlinear region of the BH curve are not accurate.

1.6 DIFFERENCES BETWEEN MAGNETIC AND ELECTRIC CIRCUITS

Whereas the similarities between magnetic and electric circuits led us to a method of analysis of magnetic circuits, recognizing the differences between the two would enable us to avoid pitfalls and errors. We consider here only differences of major consequences.

For comparison, we may refer to Fig. 1.12(a). First, the $\mathcal{F}\varphi$ characteristic of a magnetic circuit has a nonlinear form, similar to that of Fig. 1.8. On the other hand, the VI characteristic of a dc resistive circuit is generally linear. Saturation has the effect of decreasing the permeability of the material. Second, because of saturation of the main magnetic circuit, fluxes leak out via leakage paths. Such leakage fluxes are shown in Fig. 1.17. Obviously, in dc resistive circuits of the type shown in Fig. 1.12(b), the current is confined to the conductor. Third, in magnetic circuits with air gaps—especially large air gaps—fluxes tend to bulge out along the sides and around the edges of the air gap. Air gaps in magnetic circuits without and with fringing are shown in Fig. 1.18(a) and (b), respectively. Although fringing is always present, Fig. 1.12(a) shows an idealized case. Fringing has the effect of increasing the effective area of the air gap, which must be considered in the design of a magnetic circuit. The relative effect

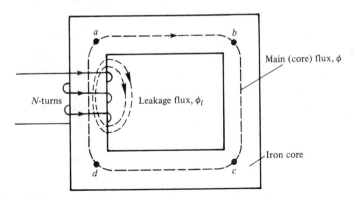

FIGURE 1-17. Leakage flux.

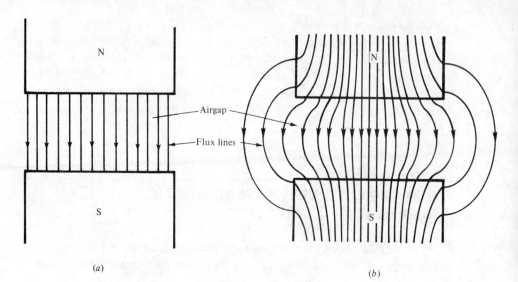

FIGURE 1-18. (a) no fringing of flux; (b) fringing of flux.

of fringing increases with the length of the air gap. However, except in very simple configurations it is almost impossible to account for fringing analytically. From experience and by quasi-analytical methods, designers have obtained fringing factors for certain special cases, where fringing must be taken into account.

Finally, a current in a resistance produces heat, or I^2R loss. However, there is no analogous $\varphi^2 \mathcal{R}$ loss in a magnetic circuit. In an electric circuit, current density is limited by heating, whereas in a magnetic circuit, flux density is limited by saturation.

1.7

AC OPERATION OF MAGNETIC CIRCUITS

By ac operation of a magnetic circuit, we mean that ac mmf's act on the circuit. How does the operation of a magnetic circuit with an ac excitation differ from that of a magnetic circuit excited by a dc mmf? The answer to this question is very important in understanding the behavior of ac machines and transformers. Also, as we will see in a later chapter, certain parts of the magnetic circuit in a dc machine (such as the armature core) operate under the influence of ac mmf's.

There are three basic differences between ac and dc magnetic circuits. Two of these differences arise from *Faraday's law* of electromagnetic induction, according to which a time-varying flux linking a coil induces an emf (or voltage) in the coil. To appreciate this law fully, let us consider a magnetic circuit having an N-turn coil wound on it, as shown in Fig. 1.19. Let a flux φ flow through the circuit. We define flux linkage, λ, by

$$\lambda = N\varphi \quad \text{Wb-turn} \tag{1.15}$$

If the flux changes with time, then, according to Faraday's law, a voltage e is induced in the N-turn coil such that

$$e = \frac{d\lambda}{dt} = N\frac{d\varphi}{dt} \quad \text{V} \tag{1.16}$$

In (1.16), d/dt implies the time rate of change. Thus the magnitude of e is simply the time rate of change of the flux linkage. The polarity of e is given by *Lenz's law*, according to which the direction of e is such as to force a current through coil, opposing the change in φ. For example, in Fig. 1.19, if the core flux φ is increasing, e will have a polarity to force a current through the coil such that the current will oppose the change (increase, in this case) in the flux producing the emf.

In formulating (1.16) we have assumed that there is no leakage flux. If the coil has negligible resistance, as in an ideal case, the induced emf e is the same as the terminal voltage v. Thus we may rewrite (1.16) as

$$v = e = N\frac{d\varphi}{dt} \tag{1.17}$$

As a consequence of (1.17), we have

$$\varphi = \frac{1}{N}\int_0^t v\,dt \tag{1.18}$$

Therefore, in an ac magnetic circuit the voltage determines the flux. This fact

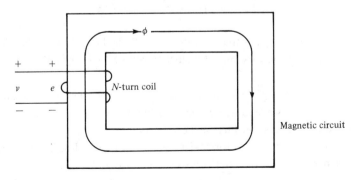

FIGURE 1-19. Faraday's law.

1.7 AC OPERATION OF MAGNETIC CIRCUITS

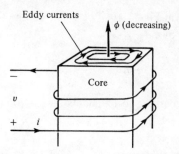

FIGURE 1-20. Eddy currents in an ac-excited core.

is one of the major differences between ac and dc magnetic circuits, and is a consequence of Faraday's law.

As we mentioned earlier, yet another difference between ac and dc magnetic circuits arises as a result of Faraday's law. Namely, a time-varying flux linking an electric circuit induces a voltage in the circuit. If the circuit is closed, a current flows through it. Keeping this fact in mind, we refer to Fig. 1.20, where the core flux is time varying. Consequently, voltages around various closed paths will be induced in the solid core. These voltages will give rise to currents, known as *eddy currents,* in the core. Eddy currents result in heating of the core and an i^2R type of loss in it. This loss is called *eddy-current loss* (nonexistent in dc magnetic circuits under steady state). This loss is given by

$$P_e = K_e f^2 B_m^2 \quad \text{W/kg} \tag{1.19}$$

where K_e is a constant depending on the material, f is the frequency of the flux variation, and B_m is the maximum value of the flux density in the core.

The third major difference between the dc and ac magnetic circuits is the presence of hysteresis in the latter (see Section 1.4). The area within the hysteresis loop is proportional to the energy loss,* as heat, per cycle; this energy loss is known as *hysteresis loss*. The hysteresis loss, P_h, is given by

$$P_h = K_h f B_m^{1.5 \text{ to } 2.5} \quad \text{W/kg} \tag{1.20}$$

where K_h is a constant, f is the frequency, and B_m is the maximum flux density. Eddy current and hysteresis losses together constitute the *core losses* (also known as *iron losses*).

To summarize, the phenomena exclusive to magnetic circuits excited by ac mmf's are hysteresis and eddy currents, leading to iron losses; and the fact that the voltage determines the flux in the core is unique to ac magnetic circuits. Figure 1.21 shows core loss for M-19 steel at 60 hertz (Hz). The exciting (or input) *VA* per pound of material is also shown in Fig. 1.21.

*We will demonstrate this fact later in this chapter.

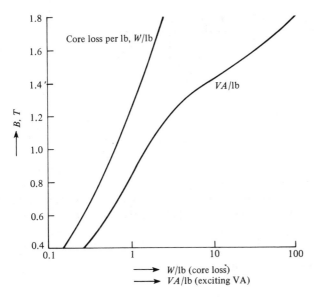

FIGURE 1-21. Core loss and exciting **VA** for M-19 steel at 60 Hz.

1.8

REDUCTION OF IRON LOSSES

To reduce eddy current loss in a magnetic circuit, its core is constructed of laminations, or thin sheets, with very thin layers of insulation between the laminations. The laminations are oriented parallel to the direction of flux, as shown in Fig. 1.22. Eddy-current loss is approximately proportional to the square of lamination thickness, which varies from 0.5 to 5 mm in most electric machines. Laminating a core increases its volume. The ratio of the volume actually occupied by the magnetic material to the total volume of the core is known as the *stacking factor*; Table 1.3 gives some values.

Because hysteresis loss is proportional to the area of the hysteresis loop, the core of a machine is made of good-quality electrical steel which has a narrow hysteresis loop.

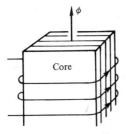

FIGURE 1-22. A portion of a laminated core.

TABLE 1.3 Stacking Factor

Lamination Thickness (mm)	Stacking Factor
0.0127	0.50
0.0254	0.75
0.0508	0.85
0.10–0.25	0.90
0.27–0.36	0.95

EXAMPLE 1.10

Consider the magnetic circuit of Fig. 1.17. Assume that the material is cast steel, having the *BH* characteristic of the form shown in Fig. 1.9. The core flux is 1 mWb and the leakage flux is 0.2 mWb. The core has a uniform cross section of 10 cm². The mean lengths *abcd* and *ad* are 75 cm and 25 cm, respectively. Assuming that the leakage flux is confined to the length *ad*, and the coil current is 5 A, determine the number of turns.

SOLUTION

Since the core flux is 1 mWb and $A = 10$ cm²,

$$B_{core} = B_{abcd} = \frac{\varphi}{A} = \frac{1 \times 10^{-3}}{10 \times 10^{-4}} = 1 \text{ T}$$

From Fig. 1.9, at $B = 1$ T,

$$H = 900 \text{ At/m}$$

Thus

$$\mathcal{F}_{abcd} = 900 \times 0.75 = 675 \text{ At}$$

Flux in the length $ad = \varphi_{core} + \varphi_{leakage} = (1 + 0.2)$ mWb, and

$$B_{ad} = \frac{1.2 \times 10^{-3}}{10 \times 10^{-4}} = 1.2 \text{ T}$$

Again from Fig. 1.9, at $B = 1.2$ T,

$$H = 1400 \text{ At/m}$$

Thus

$$\mathcal{F}_{ad} = 1400 \times 0.25 = 350 \text{ At}$$

$$\text{total mmf} = \mathcal{F}_{abcd} + \mathcal{F}_{ad} = 675 + 350 = 1025 \text{ At} = NI$$

Because $I = 5$ A, we finally obtain

$$N = \frac{1025}{5} = 205 \text{ turns}$$

EXAMPLE 1.11
For the magnetic circuit shown in Fig. 1.16, the fringing factor (defined as: effective area/actual area of cross section at the air gap) is 1.08. A 0.5-mWb flux is desired in the air gap. Assuming that $\mu_{iron} \simeq \infty$, determine the current in 100-turn coil. Compare this result with that when fringing is neglected.

SOLUTION
Since $\mathcal{F} = \mathcal{R}\varphi$ and $\mathcal{R} = l/\mu_0 A$, we have

$$\mathcal{F} = \frac{l\varphi}{\mu_0 A} = \frac{2 \times 10^{-3} \times 0.5 \times 10^{-3}}{4\pi \times 10^{-7} \times 5 \times 10^{-4} \times 1.08} = 661 \text{ At}$$

Hence

$$I = \frac{\mathcal{F}}{N} = \frac{661}{100} = 6.61 \text{ A}$$

Without fringing, we have

$$\mathcal{F} = 1.08 \times 661 = 714 \text{ At}$$

and

$$I = \frac{714}{100} = 7.14 \text{ A}$$

EXAMPLE 1.12
In a certain problem, the air-gap flux density is specified rather than the flux (as in Example 1.11). Consider the magnetic circuit of Fig. 1.16. Will the mmf required to establish the given air-gap flux density be the same, or different, if fringing is neglected, or is not neglected?

SOLUTION
From the laws of magnetic circuits we have

$$\mathcal{F} = \mathcal{R}\varphi$$

1.8 REDUCTION OF IRON LOSSES

and

$$\mathcal{R} = \frac{l}{\mu_0 A}$$

Consequently,

$$\mathcal{F} = \frac{l}{\mu_0}\frac{\varphi}{A} = \frac{lB}{\mu_0}$$

The result is independent of A. If B is specified, the mmf will remain unchanged regardless of fringing. ∎

EXAMPLE 1.13

Reconsider the magnetic circuit of Fig. 1.16. Let the circuit be made up of 0.05-mm-thick laminations of sheet metal. The BH curve is given in Fig. 1.9. Determine the current in the 100-turn coil to establish a 1-T flux density in the air gap. The 5-cm² area shown corresponds to volume occupied by the core.

SOLUTION

$$\mathcal{F}_{air} = \frac{l_{air}B}{\mu_0} = \frac{2 \times 10^{-3} \times 1}{4\pi \times 10^{-7}} = 1591 \text{ At}$$

The flux density in the core,

$$B_{core} = \frac{B_{air}}{\text{stacking factor}}$$

From Table 1.3, for 0.05-mm sheets, the stacking factor is 0.85. Therefore,

$$B_{core} = \frac{1}{0.85} = 1.17 \text{ T}$$

From Fig. 1.9 at $B = 1.17$ T,

$$H = 220 \text{ At/m}$$

Hence

$$\mathcal{F}_{core} = 220 \times 0.2 = 44 \text{ At}$$

$$\mathcal{F}_{total} = \mathcal{F}_{air} + \mathcal{F}_{core} = 1591 + 44 = 1635 \text{ At}$$

and

$$I = \frac{1635}{100} = 16.35 \text{ A}$$

EXAMPLE 1.14

A laminated core is made of M-19 steel, for which the core loss per pound is given in Fig. 1.21. Assuming that hysteresis loss is proportional to $B_m^{1.6}$, determine the constants K_e and K_h in (1.19) and (1.20), respectively.

SOLUTION

Converting (1.19) and (1.20) to pounds, we have, at 60 Hz,

$$P_e = 0.4536 \times 60^2 B_m^2 K_e = 1633 B_m^2 K_e \qquad (1)$$

and

$$P_h = 0.4536 \times 60 B_m^{1.6} K_h = 27.22 B_m^{1.6} K_h \qquad (2)$$

Hence

$$P_{\text{core}} = 1633 B_m^2 K_e + 27.22 B_m^{1.6} K_h \quad \text{W/lb} \qquad (3)$$

From Fig. 1.21, at $B_m = 1.2$ T,

$$P_{\text{core}} = 0.9 \text{ W/lb}$$

and at $B_m = 1.0$ T,

$$P_{\text{core}} = 0.65 \text{ W/lb}$$

Substituting these in (3) yields

$$2351 K_e + 36.4 K_h = 0.9 \qquad (4)$$

$$1633 K_e + 27.22 K_h = 0.65 \qquad (5)$$

From (2) and (3) we obtain

$$K_h = 1.3 \times 10^{-2}$$

and

$$K_e = 1.8 \times 10^{-4}$$

1.8 REDUCTION OF IRON LOSSES

EXAMPLE 1.15

In what proportion do hysteresis and eddy-current losses divide in M-19 steel at 60 Hz and at a maximum core flux density of 1 T?

SOLUTION

Substituting $B_m = 1$ and the values of K_e and K_h just obtained in (1) and (2) of Example 1.14, we get

$$P_e = 1633 \times 1 \times 1.8 \times 10^{-4} = 0.294 \text{ W/lb}$$

and

$$P_h = 27.22 \times 1 \times 1.3 \times 10^{-2} = 0.354 \text{ W/lb}$$

Thus

$$P_{core} = 0.294 + 0.354 = 0.65 \text{ W/lb}$$

$$\% P_e = \frac{0.294}{0.65} \times 100 = 45.37\%$$

$$\% P_h = \frac{0.354}{65} \times 100 = 54.63\%$$

1.9

INDUCTANCE

Inductance is one of the elements encountered in electric circuits. In magnetic circuits, inductance is a measure of magnetic flux linking a coil. We define inductance, L, as flux linkage per ampere; that is,

$$L = \frac{\lambda}{I} = \frac{N\varphi}{I} \quad \text{H} \tag{1.21}$$

where $\lambda = N\varphi$ is the flux linking an N-turn coil, the flux being produced by a current I. The unit of inductance is the henry (H). It would seem from (1.21) that inductance depends on the current in the coil. As we will presently see, this is not the case *except* when the magnetic circuit begins to saturate. To understand the concept of inductance further, let us refer to the magnetic circuit of Fig. 1.23. We assume that for the current I, the core has a reluctance $\mathcal{R}$. Since $\mathcal{F} = NI$, we have

$$\varphi = \frac{\mathcal{F}}{\mathcal{R}} = \frac{NI}{\mathcal{R}} \tag{1.22}$$

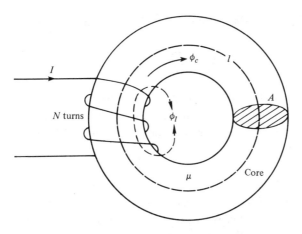

FIGURE 1-23. A magnetic circuit with an *N*-turn coil.

From (1.21) and (1.22), therefore,

$$L = \frac{N(NI)}{I \mathcal{R}} = \frac{N^2}{\mathcal{R}} \quad \text{H} \tag{1.23}$$

If the mean length of the core is l, and its area of cross section is A, then $\mathcal{R} = l/\mu A$, and (1.23) becomes

$$L = \frac{\mu A N^2}{l} \quad \text{H} \tag{1.24}$$

where μ is the permeability of the core. If μ is independent of I (that is, the core is unsaturated), then from (1.23) and (1.24) it is clear that L does not depend on I. Of course, if the core operates under saturation (in region III of Fig. 1.8), then μ, and consequently L, will depend on the I (or the degree of saturation). As a circuit element, inductance manifests itself only in time-varying situations, such as in ac circuits.

The inductance just determined, by virtue of the coil flux linkage ($N\varphi$) as produced by the coil mmf (NI), is the *self-inductance* of the coil. It is called self-inductance because the current (or mmf) and the flux linkage are associated with the same coil. Self-inductance of a coil is generally called its inductance. In the next section we make a distinction between self- and mutual inductances. For the present, let us return to Fig. 1.23. Notice that we have shown two distinct fluxes: φ_c, the core flux, and φ_l, the leakage flux. The flux φ linking the N turns is the total flux, which is the sum of φ_c and φ_l. In evaluating the effectiveness of a magnetic circuit, it is important that we have a measure of the leakage flux. The leakage flux represents a departure from the ideal case— an imperfection of the magnetic circuit, and is measured by the *leakage in-*

ductance, L_l. If φ_l is the leakage flux linking N turns, then L_l is defined by

$$L_l = \frac{N\varphi_l}{I} \quad \text{H} \tag{1.25}$$

We have discussed the significance of the total flux, φ ($= \varphi_c + \varphi_l$), and the leakage flux, φ_l. A natural question that comes to mind is: What is the significance of the core flux, φ_c? The concept of the core flux is useful in understanding mutual inductances.

EXAMPLE 1.16

A magnetic circuit of varying cross section is shown in Fig. 1.24. The BH characteristic of the iron is as given in Fig. 1.8. The data pertaining to the magnetic circuit are: $N = 100$ turns, $l_1 = 4l_2 = 0.4$ m, $A_1 = 2A_2 = 10^{-3}$ m^2, $l_g = 2$ mm, $\varphi_l = 0.01$ mWb, and $B_g = 0.6$ T. Calculate the self- and leakage inductances of the coil.

SOLUTION

From (1.21) and (1.25) it is clear that to evaluate the inductances, we have to determine the corresponding flux linkages and the operating current. Thus

$$\varphi_g = B_g A_1 = 0.6 \times 10^{-3} = 0.6 \text{ mWb}$$

$$\varphi_l = 0.01 \text{ mWb}$$

$$\varphi = \varphi_g + \varphi_l = 0.61 \text{ mWb}$$

To determine the current we must know the mmf. At $B_g = 0.6$ T,

$$H_g = \frac{0.6}{\mu_0} = 4.78 \times 10^5 \text{ A/m}$$

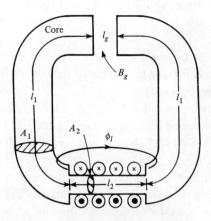

FIGURE 1-24. Example 1-16.

For air, therefore,

$$\mathcal{F}_g = H_g l_g = 4.78 \times 10^5 \times 2 \times 10^{-3} = 956 \text{ At}$$

For l_1, at $B = 0.6$ T, from Fig. 1.8, $H = 100$ A/m:

$$\mathcal{F}_{(l_1 + l_1)} = 100(0.4 + 0.4) = 80 \text{ At}$$

Since $\varphi = 0.61$ mWb (as determined above),

$$B_2 = \frac{\varphi}{A_2} = \frac{0.61}{5 \times 10^{-4}} = 1.22 \text{ T}$$

The corresponding H, from Fig. 1.8, is 410 A/m. Thus

$$\mathcal{F}_{l_2} = 410 \times 0.1 = 41 \text{ At}$$

The total mmf,

$$\mathcal{F} = NI = 956 + 80 + 41 = 1077 \text{ At}$$

or

$$I = \frac{1077}{100} = 10.77 \text{ A}$$

Therefore,

$$L = \frac{N\varphi}{I} = \frac{100 \times 0.61 \times 10^{-3}}{10.77} = 5.66 \text{ mH}$$

and

$$L_l = \frac{N\varphi_l}{I} = \frac{100 \times 0.01 \times 10^{-3}}{10.77} = 0.093 \text{ mH} \quad \blacksquare$$

1.10 MAGNETIC COUPLING OF ELECTRIC CIRCUITS: MUTUAL INDUCTANCE

In the preceding section and in Example 1.16 we mentioned core flux and leakage flux as distinct from the total flux produced by a certain mmf. Referring to Fig. 1.25, we notice that $N_1 I_1$ is an mmf resulting in a flux φ. To bring out

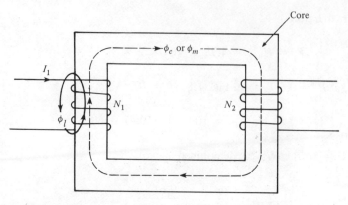

FIGURE 1-25. Mutual coupling and mutual inductance.

certain realistic conditions existing in the magnetic circuit, we decompose this total flux into φ_c, the core flux, and φ_l, the leakage flux, as shown in Fig. 1.25. Clearly, φ_l links with the coil N_1 only. However, φ_c links with the coils N_1 and N_2 both. In such a case, φ_c is called the mutual flux, φ_m, since it links with both coils. In other words, coils N_1 and N_2 are mutually coupled through the mutual flux, φ_m. The effectiveness of φ_m is measured by the *mutual inductance* between N_1 and N_2. Thus we define mutual inductance L_{12}, between two coils, as

$$L_{12} = \frac{\text{flux linking the second coil}}{\text{current in the first coil}} \qquad (1.26)$$

In terms of the symbols of Fig. 1.25, (1.26) becomes

$$L_{12} = \frac{N_2 \varphi_m}{I_1} \qquad (1.27)$$

Because φ_m is less than φ (by the amount φ_l), we express φ_m as a fraction of φ such that

$$\varphi_m = k\varphi \qquad (1.28)$$

where k is a constant that is less than 1. If $\mathcal{R}$ is the reluctance as seen by the mmf $N_1 I_1$, then

$$\varphi = \frac{N_1 I_1}{\mathcal{R}} \qquad (1.29)$$

Combining (1.27) to (1.29) yields

$$L_{12} = \frac{kN_1N_2}{\mathcal{R}} \qquad (1.30)$$

The factor k in (1.30) is known as the *coupling coefficient* and is always less than 1. As is evident from (1.28), k determines the degree of coupling between two coils.

It may be readily verified that if we define

$$L_{21} = \frac{\text{flux linking the first coil}}{\text{current in the second coil}}$$

then

$$L_{21} = L_{12} \qquad (1.31)$$

Sometimes the symbol M is used to denote the mutual inductance between two coils. However, if there are more than two mutually coupled coils present, the double-subscript notation of (1.27) or (1.31) is preferred.

The determination of mutual inductance is similar to that of self-inductance, except that for the former we must know the value of k. With the definition of k given in (1.28) and realizing that $L_1 = N_1^2/\mathcal{R}$, we may write

$$k = \frac{L_{12}}{\sqrt{L_1L_2}} = \frac{M}{\sqrt{L_1L_2}} \qquad (1.32)$$

where $0 \leq k \leq 1$.

1.11 ENERGY STORED IN MAGNETIC CIRCUITS

Consider a magnetic circuit such as the one shown in Fig. 1.12(a). Let a voltage v across the coil produce a current i through it; then in a time interval dt, the energy dW_e supplied to the coil is given by

$$dW_e = vi\,dt \qquad (1.33)$$

If the coil resistance is negligible and λ is the flux linking the coil, then, by Faraday's law, we have

$$v = \frac{d\lambda}{dt} \qquad (1.34)$$

Consequently, (1.33) becomes

$$dW_e = i\frac{d\lambda}{dt}dt = i\,d\lambda \tag{1.35}$$

Notice from Fig. 1.26 that λ depends on i. Thus, during a given interval, if λ has increased from 0 to λ while the current increases from 0 to i, the total energy supplied to the coil becomes

$$\int dW_e = W_e = \int_0^\lambda i\,d\lambda = \tfrac{1}{2} i\lambda \tag{1.36}$$

which is the shaded area shown in Fig. 1.26. The energy W_e is the electrical energy input to the coil. Since we have not included any losses in the system, W_e is stored in the coil, and is termed the *magnetic stored energy*, W_m. Various other forms of (1.36) may be obtained by using the following relationships:

$$L = \frac{\lambda}{i} = \frac{N^2}{\mathcal{R}} \tag{1.37}$$

$$\varphi = \frac{\mathcal{F}}{\mathcal{R}} = \frac{Ni}{\mathcal{R}} \tag{1.38}$$

Hence

$$W_m = \tfrac{1}{2} Li^2 = \tfrac{1}{2}\mathcal{R}\varphi^2 \tag{1.39}$$

If we assume a uniform flux density B, and a constant permeability μ, such that $\varphi = BA$ and $\mathcal{R} = l/\mu A$, then (1.39) yields

$$W_m = \frac{1}{2}\frac{l}{\mu A} B^2 A = \frac{1}{2}(lA)\frac{B^2}{\mu} \tag{1.40}$$

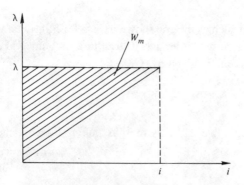

FIGURE 1-26. λ-i curve and magnetic stored energy.

Dividing both sides by lA, the volume of the material, we obtain (since $B = \mu H$)

$$W_m = \frac{W_m}{lA} = \frac{B^2}{2\mu} = \frac{1}{2}\mu H^2 \quad \text{joules/m}^3 \quad (1.41)$$

which is the energy density in the magnetic field.

1.12 EQUIVALENT CIRCUIT OF A CORE EXCITED BY AN AC MMF

As we will see in later chapters, equivalent circuits of electric machines and transformers aid in quantitatively evaluating their performances. In obtaining an electrical equivalent circuit of a machine, due attention must be paid to its magnetic circuit. The simplest magnetic circuit is a core, such as that of Fig. 1.19, with an ac mmf. We wish to represent it by an equivalent circuit. We will develop the equivalent circuit step by step as follows. First, we assume that the core has a finite and constant permeability, but no losses. We also assume that the coil has no resistance. In such an idealized case the magnetic circuit and the coil can be represented just by an inductance, L_m, as given in Fig. 1.27(a). If the coil is excited by a sinusoidal ac voltage $v = V_m \sin \omega t$, it is conventional to express L_m as an inductive reactance $X_m = \omega L_m$. This reactance is known as the *magnetizing reactance*. Thus in Fig. 1.27(a) we have X_m across which we show the terminal voltage V, equal to the induced voltage E. Note that V and E are root-mean-square (rms) values.

Next, we include the hysteresis and eddy-current losses by the resistance R_{h+e} in parallel with X_m, such that the voltage V appears across R_{h+e} also, the core losses being directly dependent on V. We thus obtain Fig. 1.27(b), where again $V = E$. Finally, we include the series resistance R, the resistance of the coil, and X_l, its *leakage reactance*, which arises from leakage fluxes. Hence we obtain

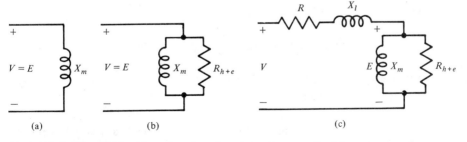

FIGURE 1-27. Equivalent circuits of an iron core excited by an ac mmf.

the electrical equivalent shown in Fig. 1.27(c), which represents an iron core excited by an ac mmf. We will use this basic equivalent circuit extensively in the study of transformers and ac machines.

QUESTIONS

1.1 A straight current-carrying conductor is located in a magnetic field and is aligned with the flux lines. Apply Ampère's law to verify that the conductor will experience no force.

1.2 Name the two sources that give rise to magnetic fields (or flux lines).

1.3 List the basic differences between a magnetic circuit and an electric circuit.

1.4 If the mmf in a magnetic circuit corresponds to an emf in an electric circuit, to what does the magnetic flux correspond?

1.5 Does a magnetic circuit have a loss similar to an I^2R loss of an electric circuit?

1.6 Why is iron preferred for magnetic circuits over materials such as wood or aluminum?

1.7 How does the operation of a magnetic circuit excited with an ac mmf differ from that of a dc-excited magnetic circuit?

1.8 What measure is taken to reduce hysteresis losses in magnetic circuits?

1.9 Why are magnetic circuits sometimes made of laminations of magnetic materials?

1.10 In choosing a material for a magnetic circuit, it is desirable to have the area of the hysteresis loop as small as possible. Why?

1.11 Explain why there will be no eddy-current loss in a dc magnetic circuit operating under steady-state conditions.

1.12 What is stacking factor? What is its value for an unlaminated magnetic core?

1.13 Distinguish between the self-inductance, mutual inductance, and leakage inductance of a coil wound on a core.

1.14 Which of the three inductances listed in Question 1.13 is a measure of imperfection of the magnetic circuit?

1.15 How is the coefficient of coupling related to the self- and mutual inductances of two mutually coupled coils?

1.16 A certain magnetic circuit is made of iron in series with an air gap. The entire circuit is of uniform cross section, but the length of the iron portion is much greater than the length of the air gap. A certain flux is established in the circuit. If the permeability of iron is much greater than the permeability of air, which portion of the magnetic circuit—iron or air— will store greater magnetic energy?

1.17 A coil is wound on a core of iron. In a certain application the core is driven to magnetic saturation. Will the inductance of the coil increase or decrease as the core saturates?

1.18 What is meant by magnetizing reactance? Compare it with leakage reactance.

1.19 Is it more desirable to have a small or a large magnetizing reactance in an electromagnetic device?

1.20 How are hysteresis and eddy-current losses represented in an equivalent circuit of a magnetic core excited by an ac mmf?

PROBLEMS

1.1 A rectangular loop with dimensions 30 cm by 20 cm is located in a uniform magnetic field of 0.2 T. Determine the magnetic flux going through the loop if its area is (a) perpendicular to the flux lines, (b) parallel to the flux lines, and (c) inclined at 45° to the flux lines.

1.2 A straight conductor is 0.5 m long and carries 15 A of current. The conductor is located in a magnetic field of 0.3 T and is inclined at an angle of 60° to the flux lines, as shown in Fig. 1P.2. Determine the force developed by the conductor.

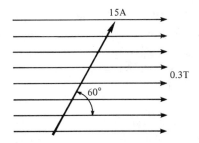

FIGURE 1P.2   PROBLEM 1.2

1.3 The core of a magnetic circuit is made in the form of a circular ring having a mean radius of 5 cm. A 200-turn coil is wound uniformly on this core. If the coil carries a 4-A current, determine (a) the mmf and (b) the magnetic field intensity in the core.

1.4 The relative permeability of the material of the ring of Problem 1.3 is 100, and the mean area of cross section of the ring is 16 cm². Calculate (a) the core flux density, (b) the core flux, (c) the reluctance, and (d) the permanence of the core.

1.5 For the ring of Problems 1.3 and 1.4, the following data for a $I\varphi$ relationship are obtained experimentally:

I (A)	0	2	4	6	8	10	12
φ (mWb)	0	0.256	0.512	0.720	0.980	1.20	1.380

where I is the coil current and φ is the core flux. Plot the BH curve for the core from the data above.

1.6 From the data of Problem 1.5, calculate the relative permeability of the core when (a) the coil current is 2 A and (b) the core flux is 1.0 mWb. (c) Determine the incremental permeability when the coil current changes from 8 A to 10 A.

1.7 A magnetic circuit has a 2-mm air gap and is made of iron having the BH characteristic shown in Fig. 1.8. The iron portion of the circuit is 25 cm long and has a uniform cross section of 8 cm². Calculate the mmf required to produce a 1-T flux density in the air gap. If a 75-turn coil is wound on the core to produce the desired flux, determine the coil current.

1.8 A magnetic circuit consisting of air and iron is shown in Fig. 1P.8. The iron has the BH characteristic shown in Fig. 1.8. The wire size of the coil is such that the coil current must not exceed 6 A. With the dimensions shown in Fig. 1P.8, find the number of turns to establish a 1.2-T flux density in the air gap.

1.9 For the magnetic circuit shown in Fig. 1P.9, assume that the iron portion is infinitely permeable. Coil N_1 carries 8 A of current. Calculate the flux densities in the gaps g_1, g_2, and g_3. Given: $g_1 = g_2 = 2$ mm, $g_3 = 4$ mm, $A_3 = 2A_1 = 2A_2 = 10$ cm², $N_1 = 60$ turns, $N_2 = 80$ turns, and $N_3 = 120$ turns. The coils N_2 and N_3 are unexcited.

1.10 Refer to the magnetic circuit of Fig. 1P.9. If all three coils are excited simultaneously such that $I_1 = 8$ A, $I_2 = 6$ A, and $I_3 = 4$ A, with the directions of currents as shown, what is the flux density in gap g_3?

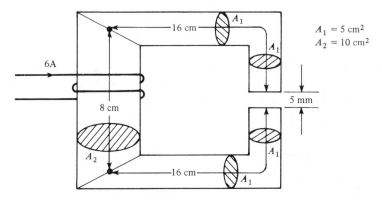

FIGURE 1P.8   PROBLEM 1.8

1.11 In Fig. 1P.9 gap g_3 is closed and only coil N_1 is excited with 8 A of current. Determine the flux densities in gaps g_1 and g_2. Next, if g_2 is also closed (other data remaining unchanged), what is the flux density in gap g_1?

1.12 Reconsider Fig. 1P.9. Using the data of Problem 1.9, evaluate the self- and mutual inductances of the three coils.

1.13 The eddy-current loss of a given solid core is 80 W at 50 Hz. What would be the eddy-current loss at 60 Hz if the core flux density remained the same at the two frequencies?

1.14 A magnetic core is made of 0.072-mm-thick laminations. The stacked core has a mean length of 35 cm and area of cross section of 12 cm². However, the area

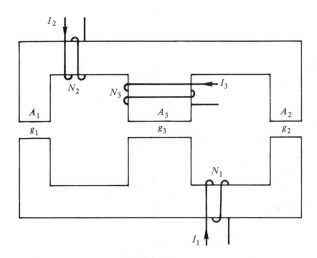

FIGURE 1P.9   PROBLEM 1.9

of cross section of the magnetic material is 10.5 cm². What is the stacking factor?

1.15 The core of a magnetic circuit is made of 0.35-mm-thick iron laminations. The magnetic circuit has an air gap in which the flux density is 0.95 T. What is the flux density in the core? If the entire magnetic circuit has a uniform cross section of 10 cm², determine the flux in the air gap and in the core.

1.16 A toroid of rectangular cross section, shown in Fig. 1P.16, has $r_1 = 8$ cm, $r_2 = 10$ cm, $a = 2$ mm, $N = 200$, and $\mu = 900\mu_0$. Assuming a uniform flux density in the core, calculate the inductance of the coil.

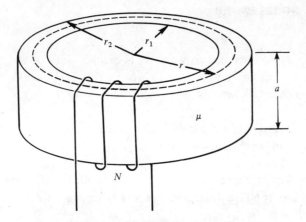

FIGURE 1P.16   PROBLEM 1.16

1.17 Evaluate the inductance of the 250-turn coil wound on the core of the magnetic circuit shown in Fig. 1P.17. Assume that the core permeability μ_i is infinity.

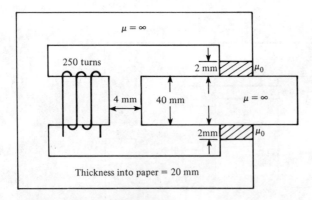

FIGURE 1P.17   PROBLEM 1.17

1.18 The magnetic circuit shown in Fig. 1P.18 is made of cast steel having the BH characteristic shown in Fig. 1.9. The air-gap flux density is 1.2 T. Because

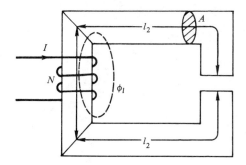

FIGURE 1P.18   PROBLEM 1.18

of saturation, there is leakage flux, φ_l, as shown in Fig. 1P.18. Given $l_2 =$ 25 cm, $l_1 =$ 10 cm, $g =$ 1 mm, $N =$ 500 turns, $\varphi_l =$ 0.08 mWb, and $A =$ 4 cm², calculate the coil current.

1.19   Calculate the leakage inductance and self-inductance of the coil shown in Fig. 1P.18.

1.20   Determine the magnetic energy stored in the entire magnetic circuit of Fig. 1P.18. Of this energy, how much is stored in the air gap?

1.21   The hysteresis loss of a certain core material is proportional to $B_m^{1.6}$. If the hysteresis loss at 1 T and 60 Hz is 160 W, determine the loss at (a) 1.2 T and 60 Hz, (b) 1 T and 50 Hz, and (c) 1.2 T and 50 Hz.

1.22   A magnetic circuit made of cast steel has a large air gap. Thus, because of fringing, the effective area of cross section at the air gap increases by 10 percent compared to the area of the core. If the air-gap flux density is 0.7 T, what is the core flux density?

CHAPTER 2

Transformers

2.1 INTRODUCTION

Among the various electromagnetic devices, the transformer has the simplest magnetic circuit. The range of applications of transformers in electrical systems extends from extremely low-voltage low-power electronic circuits to high-voltage high-power transmission lines. As a component of an electrical system, the transformer is commonly used for the following functions:

1. Changing the voltage and current levels in an electrical system, such as in a power system, where the voltage levels of power generation, transmission, and distribution are different from each other.
2. Impedance matching, as in many communication circuits where the load is matched to the line.
3. Electrical isolation, to block dc signals, to eliminate electromagnetic noise, and to provide safety in electrical appliances.

In addition to the functions listed above, transformers have other forms of applications. However, in the following we focus primarily on transformers used for voltage (or current) transformations.

2.2 TRANSFORMER CONSTRUCTION

The transformer consists of two or more electric circuits interlinking a common magnetic circuit. The windings of the transformer constitute the coupled electric circuits, whereas the *core* of the transformer provides for the magnetic circuit. Except in a certain class of transformers (known as *autotransformers,* which we discuss later in this chapter), the transformer windings are not connected electrically. Electrical power is transferred from one winding to the others electromagnetically, through the medium of the magnetic circuit. Isolation transformers must have windings that are not connected electrically.

Magnetic circuits, or cores, of transformers are made of steel laminations. Certain special types of transformers may have a powdered ferromagnetic material for its core, or the core may be of a nonmagnetic material. In the latter case the transformer is called an air-core transformer. When there are more than two windings on the core of a transformer, two of the windings usually perform identical functions. Therefore, to understand the theory and operation of a multiwinding transformer, only the relationships in two of the windings need be considered. The winding receiving power from the source is called the *primary,* and the winding supplying power to the load is called the *secondary.* A two-winding transformer having a steel core is schematically shown in Fig. 2.1. We will call the core and the windings together the electromagnetic structure of the transformer.

The construction of transformers varies greatly, depending on their applications, winding voltage and current ratings, and operating frequencies. Many electronics transformers consist of little more than the electromagnetic structure itself with a suitable means of mounting to a frame. In general, the electromagnetic structure is contained within a housing for safety and protection. In several

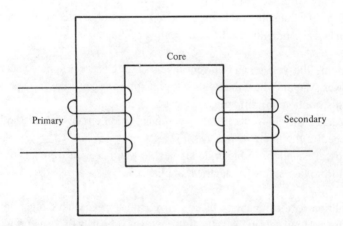

FIGURE 2-1. Model of a two-winding transformer.

types of transformers the space surrounding the electromagnetic structure is filled with an electrically insulating material to prevent damage to the windings or core and to prevent their movement or facilitate heat transfer between the electromagnetic structure and the housing. In electronics transformers a viscous insulating material called potting compound is used and, in many power transformers, a nonflammable insulating oil called transformer oil is used. Transformer oil serves the added function of improving the insulation characteristics of the transformer. In most oil-filled transformers the oil is permitted to circulate through cooling fins or tubes on the outside of the housing to improve further the heat transfer characteristics. The fins or tubes are often cooled by forced air. Figures 2.2, 2.3, and 2.4 illustrate practical transformers of various types and applications. Notice the difference between the small electronics power transformer of Fig. 2.2 and the larger power system transformers of Figs. 2.3 and 2.4. In larger transformers operating at high voltage and current levels, there are other important structural components, some of which can be seen in Fig. 2.3. Such components include porcelain bushings through which the winding leads are brought for external connection, oil pressure and temperature gauges, and internal structural supports to prevent movement of the leads or windings caused by electromagnetic forces resulting from high current levels.

The magnetic core of a transformer must be constructed in a manner to min-

FIGURE 2-2. An electronics power transformer with a core of wound strip of laminations. (Courtesy of GTE Lenkurt)

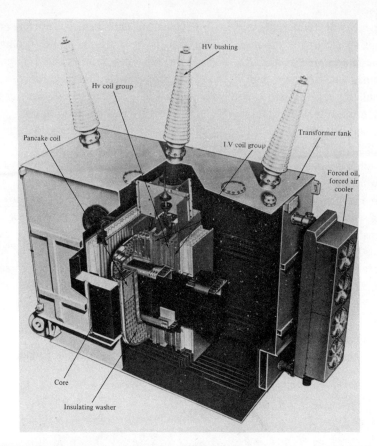

FIGURE 2-3. A 3-phase 450 MVA 345/22 kV power transformer. (Courtesy of Westinghouse Electric Corporation)

imize the magnetic losses. Power transformer cores are generally constructed from soft magnetic materials in the form of punched laminations or wound tapes. Lamination or tape thickness is a function of the transformer frequency. Pulse transformer and high-frequency electronics transformer cores are often constructed of soft ferrites. Laminations used in electronics transformers are often called "alphabet" laminations, since they are in the shape of several letters of the alphabet and designated as E, C, I, U cores, and so forth. Ferrite cores are designated by the names cup, pot, sleeve, rod, slug, and so forth, which also describe the general shape of these cores. The most common lamination materials are silicon–iron and cobalt–iron alloys. Powdered permalloy is used for many communication transformer cores.

Transformer windings are constructed of solid or stranded copper or aluminum conductors. The conductor used in electronics transformers—as well as in many small and medium-size motors and generators—is magnet wire, listed in Appendix A.

In transformers having split cores of the E, C, U, or torodial configurations,

FIGURE 2-4. A 3-phase 1300 MVA 345/24 kV power transformer. (Courtesy of Westinghouse Electric Corporation)

the windings are usually wound on an insulating spool or reel known as a bobbin. The wound bobbin is conveniently slipped over one leg of the core section, the remaining sections of core are then assembled, and the completed core assembly containing the winding is generally clamped together by means of a metal tape. The purpose of the bobbin is to provide a structural support for the winding, electrically insulate the winding from the core, and prevent abrasion or cutting of the winding at the core edges. Transformer bobbins are constructed of nylon, Teflon, and various paper and fiber products. To improve the magnetic coupling between the two windings, a bifiler winding technique is frequently used. This technique consists of laying the conductors to be used in the windings side by side and winding them on the core or bobbin simultaneously. Figure 2.5 illustrates the various stages of assembling a winding on an electronics transformer core.

The windings of large power transformers generally use conductors with heavier insulation than magnet wire insulation. The windings are assembled with much greater mechanical support, and winding layers are insulated from each other. Larger, high-powered windings are often preformed, and the transformer is assembled by stacking the laminations within the preformed coils.

FIGURE 2-5. Stages of assembly of an electronics transformer. (Courtesy of GTE Lenkurt)

2.3

TRANSFORMER CLASSIFICATION

Because of the great diversity in size, shape, and application of transformers, there has been some attempt to designate types or classes of transformers. However, the terms used to classify transformers are very loosely defined and there is much overlap among the meanings of these terms. The terms have developed more from the types of circuits in which the transformers are used than from a delineation of transformer characteristics. Since these descriptive terms are widely used in manufacturers' catalogs of transformers and since it will aid in the subsequent description of transformer characteristics and structural features, some of the terms that categorize basic transformer types are listed here.

General Application Classification

1. *Power transformers:* These are used in power distribution and transmission systems. This class has the highest power or volt-ampere ratings and the highest continuous voltage ratings. Like other ac machines, transformers are rated in volt-amperes, VA.
2. *Electronics transformers:* Transformers are of many different types and applications used in electronics circuits. Sometimes electronics transformers are considered as those transformers with ratings of 300 VA and below. A large class of electronics transformers are called ''power transformers'' and are used in supplying power to other electronic systems, which confuses the nomenclature considerably.
3. *Instrument transformers:* These are used to sense voltage or current in both electronic circuits and power systems and are called potential transformers and current transformers. The latter are series-connected devices (see Fig.

2.37) and are operated in a configuration that is, in many respects, the inverse of the conventional voltage or potential transformer.
4. *Specialty transformers:* This designation covers many styles and operating features and includes devices such as saturating, constant-voltage, constant-current, ferroresonant, and variable-tap transformers.

Classification by Frequency Range. This is probably the most significant method of classifying transformers for describing electromagnetic design features and includes the following.

1. *Power:* These are generally constant-frequency transformers that operate at the power frequencies (50, 60, 400 Hz, etc.), although other frequency components, including dc, may be present in electronics power transformers and power semiconductor transformers.
2. *Audio:* Used in many communication circuits to operate at audio frequencies.
3. *Ultra high frequency* (UHF).
4. *Wide-band:* Electronics transformers operating over a wide range of frequencies.
5. *Narrow-band:* Electronics transformers designed for a specific frequency range.
6. *Pulse:* Transformers designed for use with pulsed or chopped excitation, both in electronics and power systems applications.

Classification by Number of Windings. This has already been alluded to and includes one-winding (autotransformers), two-winding (or conventional), and multiwinding transformers, where a winding is defined as a two-terminal electrical circuit with no electrical connection to other electrical circuits.

Classification by Polyphase Connection. This classification applies principally to power transformers and refers to the method of connecting individual windings in polyphase applications. In some transformers the entire polyphase set of transformers is contained within a single housing or case, with some resulting reduction of the weight of the magnetic core. The most common connections are the wye and the delta connections of three-phase systems.

2.4

PRINCIPLE OF OPERATION: THE IDEAL TRANSFORMER

Figure 2.6 shows a two-winding ideal transformer. The transformer is ideal in the sense that its core is lossless and is infinitely permeable, has no leakage fluxes, and the windings have no losses. Absence of leakage flux implies that

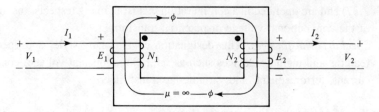

FIGURE 2-6. An ideal transformer model showing polarities and dot convention.

the entire flux links with both windings completely. In Fig. 2.6 the basic components are the *core,* the *primary winding* having N_1 turns, and the *secondary winding* having N_2 turns. If φ is the mutual (or core) flux linking N_1 and N_2, then according to Faraday's law of electromagnetic induction, emf's e_1 and e_2 are induced in N_1 and N_2 due to a time rate of change of φ such that

$$e_1 = N_1 \frac{d\varphi}{dt} \qquad (2.1)$$

and

$$e_2 = N_2 \frac{d\varphi}{dt} \qquad (2.2)$$

The direction of e_1 is such as to oppose the flux change, according to Lenz's law. The transformer being ideal, $e_1 = v_1$ (Fig. 2.6). From (2.1) and (2.2),

$$\frac{e_1}{e_2} = \frac{N_1}{N_2}$$

which may also be written in terms of rms values as

$$\frac{E_1}{E_2} = \frac{N_1}{N_2} = a \qquad (2.3)$$

where a is known as the *turns ratio.*

If the flux varies sinusoidally, such that

$$\varphi = \varphi_m \sin \omega t \qquad (2.4)$$

then from (2.1) and (2.4) the corresponding induced voltage, e, linking an N-turn winding is given by

$$e = \omega N \varphi_m \cos \omega t \qquad (2.5)$$

From (2.5) the rms value of the induced voltage is

$$E = \frac{\omega N \varphi_m}{\sqrt{2}} = 4.44 f N \varphi_m \qquad (2.6)$$

In (2.6), $f = \omega/2\pi$ is the frequency in hertz.

Equation (2.6) is known as the *emf equation* of a transformer. Without using the process of differentiation we may derive the emf equation, (2.6), as follows.

From Faraday's law, the average emf, E_a, induced in an N-turn coil linking a flux φ_m is given by

$$E_a = \frac{\lambda_m}{t_1} \qquad (2.7)$$

where λ_m is the flux linkage with the coil and t_1 is the time λ_m takes to reach its maximum value. From Chapter 1 we recall that $\lambda_m = N\varphi_m$, so that (2.7) becomes

$$E_a = \frac{N\varphi_m}{t_1} \qquad (2.8)$$

We show the emf variation (assuming it to be sinusoidal) in Fig. 2.7, from which it is clear that if f is the frequency, then $t_1 = T/4 = 1/4f$. Substituting this value of t in (2.8) yields

$$E_a = 4f\varphi_m \qquad (2.9)$$

To relate an average value of a waveform to its rms value, we obtain the *form factor* as

$$\text{form factor} = \frac{\text{rms value}}{\text{average value}} \qquad (2.10)$$

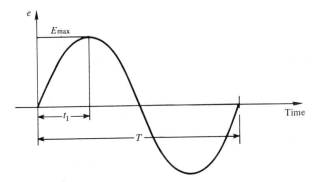

FIGURE 2-7. A sine wave, showing the relationship between t_1, the time for e to reach its maximum, and T, the period of the wave.

2.4 THE IDEAL TRANSFORMER

For a sine wave it may be shown (see Problem 2.1) that the form factor is 1.11, which implies that

$$\frac{E}{E_a} = 1.11 \tag{2.11}$$

Hence from (2.9) and (2.11) we finally get

$$E = 4.44 f N \varphi_m \tag{2.12}$$

which is identical to (2.6). In this alternative derivation, however, it is not obvious that the flux, as given by (2.4), and induced voltage, as given by (2.5), are displaced from each other by 90° in time. This 90° phase displacement is graphically illustrated in Fig. 2.8.

In conjunction with Faraday's law, we mentioned earlier in this section that the direction, or polarity, of the induced voltage is determined by Lenz's law. [See the text below (2.2).] For our purposes, we mark the polarities by using the *dot convention,* shown in Fig. 2.6. Notice that we have placed a dot at one

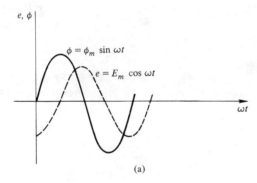

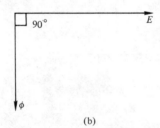

FIGURE 2-8. (a) Instantaneous relationship between the flux and the induced voltage; (b) phasor relationship between the flux and the induced voltage.

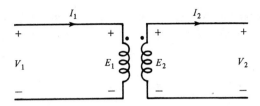

FIGURE 2-9. The dot convention of polarity marking.

terminal of each winding. This convention implies that (1) currents entering at the dotted terminals will result in mmf's that will produce fluxes in the same direction (verify this fact from Fig. 2.6); and (2) voltages from dotted to undotted terminals have the same sign. The advantage of the dot convention is that we do not have to draw the magnetic circuit, as in Fig. 2.6. Rather, Fig. 2.9, which is simpler to draw, contains the same information as that in Fig. 2.6. In fact, in Fig. 2.6 the dots are redundant and we marked the terminals there simply to introduce the convention.

2.5 VOLTAGE, CURRENT, AND IMPEDANCE TRANSFORMATIONS

As we mentioned earlier, major applications of transformers are in voltage, current, and impedance transformations, and for providing isolation (that is, eliminating direct connections between electrical circuits). The voltage transformation property, mentioned in the preceding section, of an ideal transformer is expressed as

$$\frac{V_1}{V_2} = \frac{E_1}{E_2} = a \qquad (2.13)$$

where the subscripts 1 and 2 correspond to the primary and the secondary sides, respectively. This property of a transformer enables us to interconnect transmission and distribution systems of different voltage levels in an electric power system.

For an ideal transformer, the net mmf around its magnetic circuit must be zero. In other words, we do not need any mmf (or, we need zero mmf) to establish the flux in the core of an ideal transformer. Recall from Section 1.5 the relationship between the flux φ and the mmf $\mathscr{F}$ for a magnetic circuit having a permeance $\mathscr{P}$ such that

$$\varphi = \mathscr{F}\mathscr{P} \qquad (2.14)$$

Since the core of an ideal transformer is assumed to be infinitely permeable, $\mathcal{P} = \infty$ in (2.14). Consequently, $\mathcal{F} = 0$ for φ to be finite. Alternatively, (2.14) may be written as

$$\mathcal{F} = \mathcal{R}\varphi \qquad (2.15)$$

For an infinitely permeably magnetic circuit, $\mathcal{R} = 0$. Hence for a finite φ, (2.15) requires that $\mathcal{F} = 0$. Therefore,

$$\mathcal{F} = \mathcal{F}_1 + \mathcal{F}_2 = N_1 I_1 - N_2 I_2 = 0 \qquad (2.16)$$

where I_1 and I_2 are the primary and the secondary currents, respectively. From (2.3) and (2.16) we get

$$\frac{I_2}{I_1} = \frac{N_1}{N_2} = a \qquad (2.17)$$

From (2.13) and (2.17), the impedance transformation relationships can be obtained. If Z_1 is the impedance seen at the primary terminals, then

$$Z_1 = \frac{V_1}{I_1} = \frac{E_1}{I_1} = \frac{aE_2}{I_2/a} = \frac{a^2 E_2}{I_2}$$
$$= \frac{a^2 V_2}{I_2} = a^2 Z_2 \qquad (2.18)$$

We now illustrate the discussions up to this point by the following examples.

EXAMPLE 2.1
How many turns must the primary and the secondary windings of a 220/110-V 60-Hz ideal transformer have if the core flux is not allowed to exceed 5 mWb?

SOLUTION
From the emf equation (2.6) we have

$$N = \frac{E}{4.44 f \varphi_m}$$

Consequently,

$$N_1 = \frac{220}{4.44 \times 60 \times 5 \times 10^{-3}} \approx 166 \text{ turns}$$

$$N_2 = \tfrac{1}{2} N_1 = 83 \text{ turns}$$

EXAMPLE 2.2

In a nonideal transformer, we assume that the windings have resistances. A 220/110-V 10-kVA nonideal transformer has a primary winding resistance of 0.25 ohm (Ω) and a secondary winding resistance of 0.06 Ω. Determine (a) the primary and secondary currents on a rated load; and the total resistance (b) referred to the primary and (c) referred to the secondary.

SOLUTION

(a) The transformation ratio is

$$a = \frac{220}{110} = 2$$

The primary current is

$$I_1 = \frac{10 \times 10^3}{220} = 45.45 \text{ A}$$

The secondary current is

$$I_2 = aI_1 = 2 \times 45.45 = 90.9 \text{ A}$$

(b) The secondary winding resistance referred to the primary is

$$a^2 R_2 = 2^2 \times 0.06 = 0.24 \text{ } \Omega$$

The total resistance referred to the primary is

$$R'_e = R_1 + a^2 R_2 = 0.25 + 0.24 = 0.49 \text{ } \Omega$$

(c) The primary winding resistance referred to the secondary is

$$\frac{R_1}{a^2} = \frac{0.25}{4} = 0.0625 \text{ } \Omega$$

The total resistance referred to the secondary is

$$R''_e = \frac{R_1}{a^2} + R_2 = 0.0625 + 0.06 = 0.1225 \text{ } \Omega$$

■

EXAMPLE 2.3

Determine the $I^2 R$ loss in each winding of the transformer of Example 2.2, and thus find the total $I^2 R$ loss in the two windings. Verify that the same result can be obtained by using the equivalent resistance referred to the primary winding.

SOLUTION

$$I_1^2 R_1 \text{ loss} = (45.45)^2 \times 0.25 = 516.425 \text{ W}$$

$$I_2^2 R_2 \text{ loss} = (90.9)^2 \times 0.06 = 495.768 \text{ W}$$

$$\text{total } I^2 R \text{ loss} = 1012.19 \text{ W}$$

For the equivalent resistance, R_e',

$$I_1^2 R_e' = (45.45)^2 \times 0.49 = 1012.19 \text{ W}$$

which is consistent with the preceding result. ∎

EXAMPLE 2.4
A 220/110-V 60-Hz ideal transformer has 166 turns on its primary. What is the instantaneous flux?

SOLUTION
From (2.6) we have

$$\varphi_m = \frac{E}{4.44 fN} = \frac{220}{4.44 \times 60 \times 166} = 4.975 \text{ mWb}$$

Since $f = 60$ Hz,

$$\omega = 2\pi f = 2\pi \times 60 = 377 \text{ radians/second (rad/s)}$$

Hence from (2.4),

$$\varphi = 4.975 \sin 377t \quad \text{mWb}$$

In general, however, we observe that the flux and voltage are related by (2.1), or

$$\varphi = \frac{1}{N} \int v \, dt \qquad (2.19)$$

∎

EXAMPLE 2.5
The load on an ideal transformer having a turns ratio of 5 is 10 Ω. What is the ohmic value of impedance as seen (or measured) at the primary terminals?

SOLUTION

This problem illustrates a direct application of (2.18), in which we substitute $a = 5$ and $Z_2 = Z_{load} = 10\ \Omega$. Hence

$$Z_1 = 5^2 \times 10 = 250\ \Omega$$

The problem shows the impedance transformation property of the transformer. In this particular case, a 10-Ω impedance at the secondary would appear as a 250-Ω impedance at the primary. ∎

2.6 THE NONIDEAL TRANSFORMER

A nonideal (or an actual) transformer differs from an ideal transformer in that the former has hysteresis and eddy-current (or core) losses, and has resistive (I^2R) losses in its primary and secondary windings, as indicated in Examples 2.2 and 2.3 Furthermore, the core of a nonideal transformer is not perfectly permeable, and the transformer core requires a finite mmf for its magnetization. Also, not all fluxes link with the primary and the secondary windings simultaneously because of leakage. Referring to Fig. 2.10, we observe that R_1 and R_2 are the respective resistances of the primary and secondary windings. The flux φ_c, which replaces the flux φ of Fig. 2.6, is called the *core flux* or *mutual flux*, as it links both the primary and the secondary windings. The primary and the secondary leakage fluxes are shown as φ_{l1} and φ_{l2}, respectively. Thus in Fig. 2.10 we have accounted for all the imperfections listed above, except the core losses. We will include the core losses as well as the rest of the imperfections in the equivalent circuit of a nonideal transformer. This circuit is also known as the *exact equivalent circuit*, as it differs from the idealized equivalent circuit and the various approximate equivalent circuits.

The development of the equivalent circuit follows from Figs. 1.25 and 1.27.

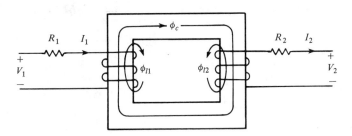

FIGURE 2-10. A non-ideal transformer showing the leakage fluxes and the winding resistances.

To recall from Chapter 1, the leakage fluxes φ_{l1} and φ_{l2} give rise to the leakage reactances, X_1 and X_2, respectively. The mutual flux φ_c is represented by the magnetizing reactance X_m and the core losses are represented by a resistance, R_c, in parallel with X_m, as was done in Fig. 1.27. Consequently, the equivalent circuit of an ideal transformer, shown in Fig. 2.11(a), modifies to that shown in Fig. 2.11(b), which is the exact equivalent circuit of a nonideal transformer. In Fig. 2.11(b), notice that the circuit components denoting the imperfections of the transformer are coupled by an ideal transformer (of proper turns ratio). This ideal transformer may be removed and the entire equivalent circuit may be referred either to the primary or to the secondary of the transformer by using its transformation properties, as given by (2.13), (2.17), and (2.18).

Referring to the primary side, according to (2.18), an impedance Z_2, on the secondary side, will appear as $a^2 Z_2$ on the primary side. In accordance with (2.17), a current I_2 on the secondary side corresponds to a current I_2/a on the primary side. Finally, to eliminate the ideal transformer from Fig. 2.11(b), and refer the secondary quantities to the primary, we use (2.13) to replace E_2 by $aE_2 = E_1$. The circuit then becomes as shown in Fig. 2.12. Of course, the primary quantities (V_1, I_1, R_1, and X_1) remain unchanged.

At this point, let us pose two questions. First, why is it necessary to obtain an equivalent circuit referred to the primary (or secondary)? Second, what is the physical meaning of referring secondary quantities to the primary (and vice versa)? In answer to the first question, by referring the entire circuit to the

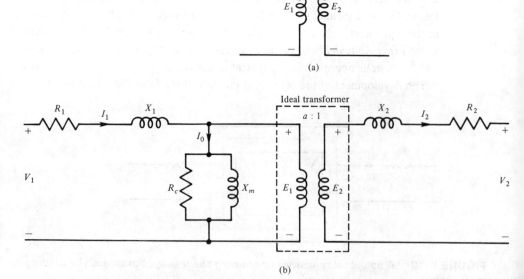

FIGURE 2-11. Equivalent circuits of (a) ideal and (b) non-ideal transformers.

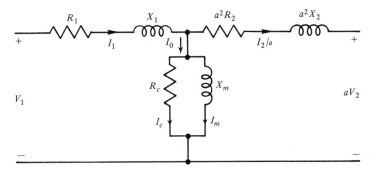

FIGURE 2-12. Equivalent circuit referred to primary.

primary, we have eliminated the ideal transformer. Thus the transformer can be represented by an *RL* circuit (Fig. 2.12). Such a representation involves a simpler circuit analysis than that of the circuit of Fig. 2.11(b). Furthermore, an *RL* circuit of the type shown in Fig. 2.12 can be approximated by simpler circuits in a straightforward manner, as we shall see later. In other words, referring the equivalent circuit (either exact or approximate) to one side (either primary or secondary) of the transformer results in a simplification of the analysis. Turning to the second question, referring the secondary impedance to the primary side implies that the real and reactive powers in an impedance Z_2 through which the secondary current I_2 flows remain the same when the primary current I_1 flows through an equivalent impedance Z_2'. This restriction ensures that the performance of a transformer as calculated from a circuit referred to the primary or from one referred to the secondary remains the same as the results obtained from the circuit of Fig. 2.11(b). In simple terms, the circuits of Figs. 2.12 and 2.11(b) must give the same results.

To obtain the equivalent circuit referred to the secondary, we again use (2.13), (2.17), and (2.18) Thus we get the equivalent circuit shown in Fig. 2.13. It is left as an exercise for the student to verify that the circuit of Fig. 2.13 is indeed an equivalent circuit of a transformer referred to its secondary.

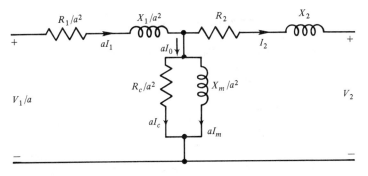

FIGURE 2-13. Equivalent circuit referred to secondary.

2.6 THE NONIDEAL TRANSFORMER

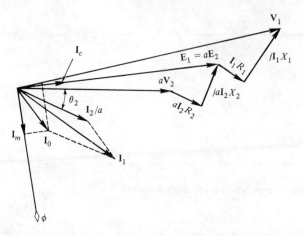

FIGURE 2-14. Phasor diagram corresponding to Fig. 2-12.

A phasor diagram for the circuit of Fig. 2.12, for lagging power factor, is shown in Fig. 2.14. In Figs. 2.12 to 2.14, the various symbols* are

$a \equiv$ turns ratio

$\mathbf{E}_1 \equiv$ primary induced voltage

$\mathbf{E}_2 \equiv$ secondary induced voltage

$\mathbf{V}_1 \equiv$ primary terminal voltage

$\mathbf{V}_2 \equiv$ secondary terminal voltage

$\mathbf{I}_1 \equiv$ primary current

$\mathbf{I}_2 \equiv$ secondary current

$\mathbf{I}_0 \equiv$ no-load (primary) current

$R_1 \equiv$ resistance of the primary winding

$R_2 \equiv$ resistance of the secondary winding

$X_1 \equiv$ primary leakage reactance

$X_2 \equiv$ secondary leakage reactance

$\mathbf{I}_m, X_m \equiv$ magnetizing current and reactance

$\mathbf{I}_c, R_c \equiv$ current and resistance accounting for the core losses

*Boldface letters indicate phasor quantities. Italic symbols are used for magnitudes only.

2.7 PERFORMANCE CHARACTERISTICS OF A TRANSFORMER

The major use of the equivalent circuit of a transformer is in determining its characteristics. The characteristics of most interest to power engineers are voltage regulation and efficiency. *Voltage regulation* is a measure of the change in the terminal voltage of the transformer with load. From Fig. 2.14 it is clear that the terminal voltage V_1 is load dependent. Specifically, we define voltage regulation as:

$$\text{percent regulation} = \frac{V_{\text{no load}} - V_{\text{load}}}{V_{\text{load}}} \times 100 \qquad (2.20)$$

With reference to Fig. 2.14, we may rewrite (2.20) as

$$\text{percent regulation} = \frac{V_1 - aV_2}{aV_2} \times 100 \qquad (2.21)$$

There are two kinds of efficiencies of transformers of interest to us, known as *power efficiency* and *energy efficiency*. These are defined as follows:

$$\text{power efficiency} = \frac{\text{output power}}{\text{input power}} \qquad (2.22)$$

$$\text{energy efficiency} = \frac{\text{output energy for a given period}}{\text{input energy for the same period}} \qquad (2.23)$$

Generally, energy efficiency is taken over a 24-hour (h) period and is called *all-day efficiency*. In such a case (2.23) becomes

$$\text{all-day efficiency} = \frac{\text{output for 24 h}}{\text{input for 24 h}} \qquad (2.24)$$

For our purposes, we will use the term *efficiency* to mean power efficiency from now on. It is clear from our discussion of nonideal transformers that the output power is less than the input power because of losses, the losses being I^2R losses in the windings, and hysteresis and eddy-current losses in the core. Thus, in terms of these losses, (2.22) may be more meaningfully expressed as

$$\text{efficiency} = \frac{\text{input power} - \text{losses}}{\text{input power}}$$

$$= \frac{\text{output power}}{\text{output power} + \text{losses}} \tag{2.25}$$

$$= \frac{\text{output power}}{\text{output power} + I^2R \text{ loss} + \text{core loss}}$$

Obviously, I^2R loss is load dependent, whereas the core loss is constant and independent of the load on the transformer. The next examples show voltage regulation and efficiency calculations of a transformer.

EXAMPLE 2.6
A 150-kVA 2400/240-V transformer has the following parameters: $R_1 = 0.2\ \Omega$, $R_2 = 0.002\ \Omega$, $X_1 = 0.45\ \Omega$, $X_2 = 0.0045\ \Omega$, $R_c = 10{,}000\ \Omega$, and $X_m = 1550\ \Omega$, where the symbols are shown in Fig. 2.11(b). Refer the circuit to the primary. From this circuit, calculate the voltage regulation of the transformer at rated load with a 0.8 lagging power factor.

SOLUTION
The circuit referred to the primary is shown in Fig. 2.12. From the data given, we have $V_2 = 240$ V, $a = 10$, and $\theta_2 = \cos^{-1} 0.8 = 36.8°$.

$$I_2 = \frac{150 \times 10^3}{240} = 625\ \text{A}$$

$$\frac{\mathbf{I}_2}{a} = 62.5\ \underline{/-36.8} = 50 - j37.5\ \text{A}$$

$$a\mathbf{V}_2 = 2400\ \underline{/0°} = 2400 + j0\ \text{V}$$

$$a^2 R_2 = 0.2\ \Omega \quad \text{and} \quad a^2 X_2 = 0.45\ \Omega$$

Hence

$$\mathbf{E}_1 = (2400 + j0) + (50 - j37.5)(0.2 + j0.45)$$

$$= 2427 + j15 = 2427\ \underline{/0.35°}\ \text{V}$$

$$\mathbf{I}_m = \frac{2427\ \underline{/0.35°}}{1550\ \underline{/90°}} = 1.56\ \underline{/-89.65} = 0.0095 - j1.56\ \text{A}$$

$$\mathbf{I}_c = \frac{2427 + j15}{10{,}000} \approx 0.2427 + j0\ \text{A}$$

$$\mathbf{I}_0 = \mathbf{I}_c + \mathbf{I}_m = 0.25 - j1.56\ \text{A}$$

$$I_1 = \frac{I_0 + I_2}{a} = 50.25 - j39.06 = 63.65 \underline{/-37.85}$$

$$V_1 = (2427 + j15) + (50.25 - j39.06)(0.2 + j0.45)$$

$$= 2455 + j30 = 2455 \underline{/0.7°} \text{ V}$$

$$\text{percent regulation} = \frac{V_1 - aV_2}{aV_2} \times 100$$

$$= \frac{2455 - 2400}{2400} \times 100 = 2.3\%$$

It may be verified that for a leading power factor load the voltage regulation can be negative (see Problem 2.7). ∎

EXAMPLE 2.7
Determine the efficiency of the transformer of Example 2.6 operating on rated load and a 0.8 lagging power factor.

SOLUTION

$$\text{output} = 150 \times 0.8 = 120 \text{ kW}$$

$$\text{losses} = I_1^2 R_1 + I_c^2 R_c + I_2^2 R_2$$

$$= (63.65)^2 \times 0.2 + (0.2427)^2 \times 10{,}000 + (625)^2 \times 0.002$$

$$= 2.18 \text{ kW}$$

$$\text{input} = 120 + 2.18 = 122.18 \text{ kW}$$

$$\text{efficiency} = \frac{120}{122.18} = 98.2\%$$

∎

EXAMPLE 2.8
The transformer of Example 2.6 operates on full-load 0.8 lagging power factor for 12 h, on no-load for 4 h, and on half-full load unity power factor for 8 h. Calculate the all-day efficiency.

SOLUTION

$$\text{output for 24 h} = (150 \times 0.8 \times 12) + (0 \times 4) + (150 \times \tfrac{1}{2} \times 8)$$

$$= 2040 \text{ kWh}$$

The losses for 24 h are

$$\text{core loss} = (0.2427)^2 \times 10{,}000 \times 24 = 14.14 \text{ kWh}$$

I^2R loss on full load for 12 h

$$= 12[(63.65)^2 \times 0.2 + (625)^2 \times 0.002] = 19.1 \text{ kWh}$$

I^2R loss on half-full load for 8 h

$$= 8\left[\left(\frac{63.65}{2}\right)^2 \times 0.2 + \left(\frac{625}{2}\right) \times 0.002\right] = 3.18 \text{ kWh}$$

$$\text{total losses for 24 h} = 14.14 + 19.1 + 3.18 = 36.42$$

$$\text{input for 24 h} = 2040 + 35.42 = 2076.42 \text{ kWh}$$

$$\text{all-day efficiency} = \frac{2040}{2076.42} = 98.2\%$$

2.8

APPROXIMATE EQUIVALENT CIRCUITS

Let us review the results of Example 2.6. Notice that the no-load current $I_0 = 1.6$ A, whereas the load current $I_1 = 63.6$ A. Also, $E_1 = 2427$ V and $V_1 = 2455$ V. In a well-designed commercial transformer, $I_0 \ll I_1$ and $E_1 \simeq V_1$. As a result, we can move the shunt (no-load) circuit to the primary terminals (extreme left, as in Fig. 2.15). The circuit shown in Fig. 2.15 is known as the *approximate equivalent circuit* of the transformer. The principal advantage of using the approximate equivalent is the reduction in the amount of complex arithmetic computations. In order to compare the calculations involved in the approximate equivalent circuit, let us solve the problem of Example 2.6 again as follows.

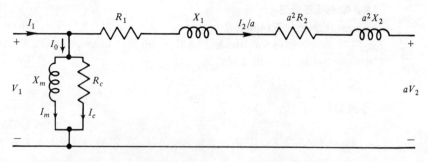

FIGURE 2-15. An approximate equivalent circuit referred to the primary.

EXAMPLE 2.9

An approximate equivalent circuit of a transformer is shown in Fig. 2.15. Using this circuit, repeat the calculations of Example 2.6 and compare the results. Draw a phasor diagram showing all the voltages and currents of the circuit of Fig. 2.15.

SOLUTION

Using Example 2.6, we have

$$a\mathbf{V}_2 = 2400 \underline{/0°} \quad \text{V}$$

$$\frac{\mathbf{I}_2}{a} = 50 - j37.5 \quad \text{A}$$

$$R_1 + a^2 R_2 = 0.4 \ \Omega$$

$$X_1 + a^2 X_2 = 0.9 \ \Omega$$

Hence

$$\mathbf{V}_1 = (2400 + j0) + (50 - j37.5)(0.4 + j0.9)$$

$$= 2453 + j30 = 2453 \underline{/0.7°} \text{ V}$$

$$\mathbf{I}_c = \frac{2453 \underline{/0.7°}}{10 \times 10^3} = 0.2453 \underline{/0.7°} \text{ A}$$

$$\mathbf{I}_m = \frac{2453 \underline{/0.7°}}{1550 \underline{/90°}} = 1.58 \underline{/-89.3°} \text{ A}$$

$$\mathbf{I}_0 = 0.2453 - j1.58 \text{ A}$$

$$\mathbf{I}_1 = 50.25 - j39.08 = 63.66 \underline{/-37.9°} \text{ A}$$

The phasor diagram is shown in Fig. 2.16.

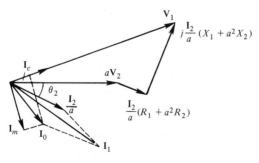

FIGURE 2-16. Phasor diagram corresponding to the equivalent circuit of Fig. 2.15.

$$\text{percent regulation} = \frac{2453 - 2400}{2400} \times 100 = 2.2\%$$

$$\text{efficiency} = \frac{120 \times 10^3}{120 \times 10^3 + (63.66)^2(0.4) + (0.2453)^2(10 \times 10^3)}$$

$$= 0.982 = 98.2\%$$

Notice that the approximate circuit yields results that are sufficiently accurate. ∎

The circuit shown in Fig. 2.15 is the approximate equivalent of the transformer referred to the primary. As with the exact equivalent circuit, we may obtain the approximate equivalent circuit referred to either of the two windings. The next example illustrates the procedure.

EXAMPLE 2.10
The ohmic values of the circuit parameters of a transformer, having a turns ratio of 5, are $R_1 = 0.5 \; \Omega$; $R_2 = 0.021 \; \Omega$; $X_1 = 3.2 \; \Omega$; $X_2 = 0.12 \; \Omega$; $R_c = 350 \; \Omega$, referred to the primary; and $X_m = 98 \; \Omega$, referred to the primary. Draw the approximate equivalent circuits of the transformer, referred to (a) the primary and (b) the secondary. Show the numerical values of the circuit parameters.

SOLUTION
The circuits are shown in Fig. 2.17(a) and (b), respectively. The calculations are as follows:

(a) $R' \equiv R_1 + a^2 R_2 = 0.5 + (5)^2(0.021) = 1.025 \; \Omega$

$X' \equiv X_1 + a^2 X_2 = 3.2 + (5)^2(0.12) = 6.2 \; \Omega$

$R'_c = 350 \; \Omega$

$X'_m = 98 \; \Omega$

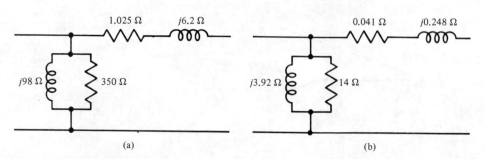

FIGURE 2-17. Example 2.10.

(b) $R'' \equiv \dfrac{R_1}{a^2} + R_2 = \dfrac{0.5}{25} + 0.021 = 0.041 \ \Omega$

$X'' \equiv \dfrac{X_1}{a^2} + X_2 = \dfrac{3.2}{25} + 0.12 = 0.248 \ \Omega$

$R''_c = \dfrac{350}{25} = 14 \ \Omega$

$X''_m = \dfrac{98}{25} = 3.92 \ \Omega$

2.9 EQUIVALENT CIRCUITS FROM TEST DATA

The preceding two sections have clearly indicated the utility of transformer equivalent circuits. The parameters—the resistances and the reactances—of the equivalent circuits may be obtained from the following tests.

Open-Circuit (or No-Load) Test. In this test, one winding is open-circuited and a voltage—usually, rated voltage at rated frequency—is applied to the other winding. The voltage, current, and power at the terminals of this winding are measured. The open-circuit voltage of the second winding is also measured, and from this measurement a check on the turns ratio can be obtained. It is usually convenient to apply the test voltage to the winding that has a voltage rating equal to that of the available power source. This means that in step-up voltage transformers, the open-circuit voltage of the second winding will be higher than the applied voltage, sometimes much higher. Care must be exercised in guarding the terminals of this winding to ensure safety for test personnel and to prevent these terminals from getting close to other electrical circuits, instrumentation, grounds, and so forth.

In presenting the no-load parameters obtainable from test data, it is assumed that voltage is applied to the primary and that the secondary is open-circuited. The no-load power loss is equal to the wattmeter reading in this test; core loss is found by subtracting the ohmic loss in the primary, which is usually small and may be neglected in some cases. Thus if P_0, I_0, and V_0 are the input power, current, and voltage, respectively, the core loss, P_c, is given by

$$P_c = P_0 - I_0^2 R_1$$

The primary induced voltage is expressed in phasor form by

$$\mathbf{E}_1 = V_0 \underline{/0°} - (I_0 \underline{/\theta_0})(R_1 + jX_1)$$

where $\theta_0 \equiv$ no-load power-factor angle $= \cos^{-1}(P_0/V_0I_0) < 0$. Other circuit quantities are found from

$$R_c = \frac{E_1^2}{P_c}$$

$$I_c = \frac{P_c}{E_1}$$

$$I_m = \sqrt{I_0^2 - I_c^2}$$

$$X_m = \frac{E_1}{I_m}$$

$$a \approx \frac{V_0}{E_2}$$

The equivalent circuit for the open-circuit test is shown in Fig. 2.18.

Short-Circuit Test. In this test, one winding is short-circuited across its terminals, and a reduced voltage is applied to the other winding. This reduced voltage is of such a magnitude as to cause a specific value of current—usually, rated current—to flow in the short-circuited winding. Again, the choice of the winding to be short-circuited is usually determined by the measuring equipment available for us in the test. However, care must be taken to note which winding is short-circuited, for this determines the reference winding for expressing the impedance components obtained by this test. Let the secondary be short-circuited and the reduced voltage be applied to the primary.

With a very low voltage applied to the primary winding, the core-loss current and magnetizing current become very small, and the equivalent circuit reduces to that of Fig. 2.19. Thus if P_s, I_s, and V_s are, respectively, the input power, current, and voltage under short circuit, then, referred to the primary,

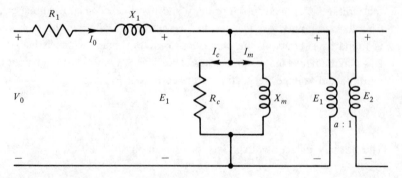

FIGURE 2-18. Equivalent circuit for open circuit test.

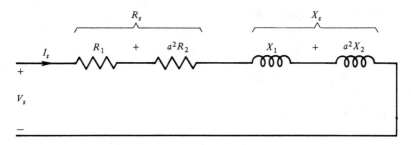

FIGURE 2-19. Equivalent circuit for short circuit test.

$$Z_s = \frac{V_s}{I_s}$$

$$R_1 + a^2 R_2 \equiv R_s = \frac{P_s}{I_s^2} \quad (2.26)$$

$$X_1 + a^2 X_2 \equiv X_s = \sqrt{Z_s^2 - R_s^2} \quad (2.27)$$

Given R_1 and a, R_2 can be found from (2.26). In (2.27) it is usually assumed that the leakage reactance is divided equally between the primary and the secondary; that is,

$$X_1 = a^2 X_2 = \tfrac{1}{2} X_s$$

Notice that the open-circuit test is performed with the instrumentation on the low-voltage side, whereas the instrumentation is on the high-voltage side for the short-circuit test. Thus we have R_c and X_m referred to the secondary (low-voltage side) and R_1, R_2, X_1, and X_2 referred to the primary (high-voltage side). The equivalent circuit, however, must be referred to one side only. The next two examples show how we interpret the data obtained from the two tests.

EXAMPLE 2.11

A certain transformer, with its secondary open, takes 80 W of power at 120 V and 1.4 A. The primary winding resistance is 0.25 Ω and the leakage reactance is 1.2 Ω. Evaluate the magnetizing reactance, X_m, and the core-loss equivalent resistance, R_c.

SOLUTION
The no-load power factor angle is

$$\theta_0 = \cos^{-1} \frac{80}{1.4 \times 120} = -61.6°$$

2.9 EQUIVALENT CIRCUITS FROM TEST DATA

The primary induced voltage is

$$E_1 = 120 \underline{/0} + 1.4 \underline{/-61.6}\,(0.25 + j1.25)$$
$$\simeq 118.29 \text{ V}$$

Therefore,

$$R_c = \frac{(118.29)^2}{80 - (1.4)^2(0.25)} = 176 \text{ }\Omega$$

$$I_c = \frac{118.29}{176} = 0.672 \text{ A}$$

$$I_m = \sqrt{(1.4)^2 - (0.672)^2} = 1.228 \text{ A}$$

$$X_m = \frac{118.29}{1.228} = 96.3 \text{ }\Omega$$

∎

EXAMPLE 2.12

The results of open-circuit and short-circuit tests on a 25-kVA 440/220-V 60-Hz transformer are as follows:

Open-circuit test: Primary open-circuited, with instrumentation on the low-voltage side. Input voltage, 200 V; input current, 9.6 A; input power, 710 W.

Short-circuit test: Secondary short-circuited, with instrumentation on the high-voltage side. Input voltage, 42 V; input current, 57 A; input power, 1030 W. Obtain the parameters of the exact equivalent circuit (Fig. 2.12), referred to the high-voltage side. Assume that $R_1 = a^2 R_2$ and $X_1 = a^2 X_2$.

SOLUTION

From the short-circuit test:

$$Z_{s1} = \frac{42}{57} = 0.737 \text{ }\Omega$$

$$R_{s1} = \frac{1030}{(57)^2} = 0.317 \text{ }\Omega$$

$$X_{s1} = \sqrt{(0.737)^2 - (0.317)^2} = 0.665 \text{ }\Omega$$

Consequently,

$$R_1 = a^2 R_2 = 0.158 \text{ }\Omega \qquad R_2 = 0.0395 \text{ }\Omega$$
$$X_1 = a^2 X_2 = 0.333 \text{ }\Omega \qquad X_2 = 0.0832 \text{ }\Omega$$

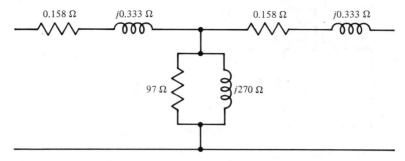

FIGURE 2-20. Example 2.12.

From the open-circuit test:

$$\theta_0 = \cos^{-1} \frac{710}{(9.6)(220)} = \cos^{-1} 0.336 = -70°$$

$$E_2 = 220 \underline{/0°} - (9.6 \underline{/-70°})(0.0395 + j0.0832) \approx 219 \underline{/0°} \text{ V}$$

$$P_{c2} = 710 - (9.6)^2(0.0395) \approx 710 \text{ W}$$

$$R_{c2} = \frac{(219)^2}{710} = 67.5 \ \Omega$$

$$I_{c2} = \frac{219}{67.5} = 3.24 \text{ A}$$

$$I_{m2} = \sqrt{(9.6)^2 - (3.24)^2} = 9.03 \text{ A}$$

$$X_{m2} = \frac{219}{9.03} = 24.24 \ \Omega$$

$$X_{m1} = a^2 X_{m2} = 97 \ \Omega$$

$$R_{c1} = a^2 R_{c2} = 270 \ \Omega$$

The equivalent thus obtained is shown in Fig. 2.20. ∎

2.10

TRANSFORMER POLARITY

Polarities of a transformer identify the relative directions of induced voltages in the two windings. The polarities result from the relative directions in which the two windings are wound on the core. For operating transformers in parallel, and for various transformer connections, it is necessary that we know the relative

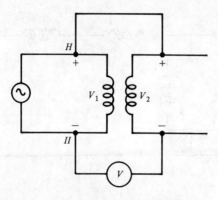

FIGURE 2-21. Polarity test on a transformer.

polarities. Polarities can be checked by a simple test, requiring only voltage measurements with the transformer on no-load. In this test, rated voltage is applied to one winding, and an electrical connection is made between one terminal from one winding and one from the other, as shown in Fig. 2.21. The voltage across the two remaining terminals (one from each winding) is then measured. If this measured voltage is *larger* than the input test voltage, the polarity is *additive*; if smaller, the polarity is *subtractive*.

A standard method of marking transformer terminals is as follows. The high-voltage terminals are marked H1, H2, H3, . . . , with H1 on the right-hand side of the case when facing the high-voltage side. The low-voltage terminals are designated X1, X2, X3, . . . , and X1 may be on either side, adjacent to H1 or diagonally opposite. The two possible locations of X1 with respect to H1 for additive and subtractive polarities are shown in Fig. 2.22. The numbers must be so arranged that the voltage difference between any two leads of the same set, taken in order from smaller to larger numbers, must be of the same sign as that between any other pair of the set taken in the same order. Furthermore, when the voltage is directed from H1 to H2, it must simultaneously be directed from X1 to X2. Additive polarities are required by the American National Standards Institute (ANSI) in large (>200 kVA) high-voltage (>8660 V) power trans-

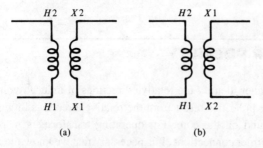

FIGURE 2-22. (a) Subtractive and (b) additive polarities of a transformer.

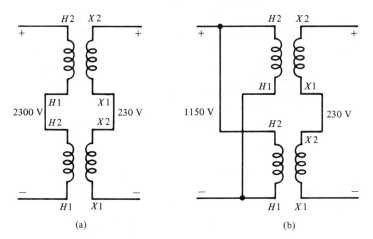

FIGURE 2-23. Example 2.13.

formers. Small transformers have subtractive polarities (which reduce voltage stress between adjacent loads).

Knowing the polarities of a transformer aids in obtaining a combination of several voltages from the transformer. Certain single-phase connections of transformers are illustrated by the next example.

EXAMPLE 2.13

Two 1150/115-V transformers are given. Using appropriate polarity markings, show the interconnections of these transformers for (a) 2300/230-V operation and (b) 1150/230-V operation.

SOLUTION

We mark the high-voltage terminals by H1 and H2 and the low-voltage terminals by X1 and X2 (as in Fig. 2.22). According to the ANSI convention, H1 is marked on the terminal on the right-hand side of the transformer case (or housing) when facing the high-voltage side. With this nomenclature, the desired connections are shown in Fig. 2.23. ∎

2.11

TRANSFORMERS IN PARALLEL

Transformers are connected in parallel to share loads exceeding the capacity of an individual transformer. Let us consider two transformers operating in parallel (with proper polarities) and supplying a common load, as given in Fig. 2.24. Neglecting the core losses and magnetizing currents, the equivalent circuit be-

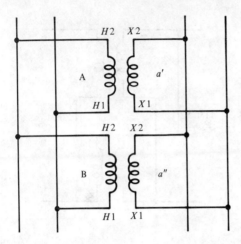

FIGURE 2-24. Two transformers in parallel.

comes as shown in Fig. 2.25. If the transformers have a' and a'' as their respective turns ratio, then from Fig. 2.25 we have

$$\mathbf{V}_1 = \mathbf{V}_1' = a'\mathbf{V}_t + \mathbf{I}_1'\mathbf{Z}_e' \tag{2.28}$$

$$\mathbf{V}_1 = \mathbf{V}_2'' = a''\mathbf{V}_t + \mathbf{I}_1''\mathbf{Z}_e'' \tag{2.29}$$

$$\mathbf{I}_1 = \mathbf{I}_1' + \mathbf{I}_1'' \tag{2.30}$$

Subtracting (2.29) from (2.28) and solving with (2.30) yields

$$\mathbf{I}_1' = \frac{-\mathbf{V}_t(a' - a'') + \mathbf{I}_1\mathbf{Z}_e''}{\mathbf{Z}_e' + \mathbf{Z}_e''} \tag{2.31}$$

$$\mathbf{I}_1'' = \frac{\mathbf{V}_t(a' - a'') + \mathbf{I}_1\mathbf{Z}_e'}{\mathbf{Z}_e' + \mathbf{Z}_e''} \tag{2.32}$$

Notice from (2.31) and (2.32) that the current through any one of the transformers consists of two components. One component is proportional to I_1 or the

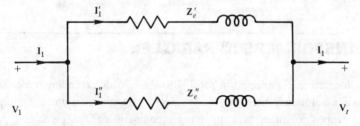

FIGURE 2-25. Equivalent circuit of transformers in parallel.

load, whereas the second component varies with $V_t(a' - a'')$. The latter component is constant, and gives rise to the local circulating current between the two transformers. Clearly, with transformers having an equal turns ratio, $a' = a''$, there will be no circulating current.

EXAMPLE 2.14
A 2300/230-V transformer has an equivalent series impedance of $1.84 \underline{/84.2°}$ Ω referred to the primary. Another transformer rated at 2300/230-V has an equivalent impedance of $0.77 \underline{/82.5°}$ Ω. The two transformers are connected in parallel and supply a 230-V 400-kVA 0.8 lagging power factor load. Determine the power supplied by each transformer.

SOLUTION
This problem involves an application of (2.31) and (2.32). However, since the load is specified on the secondary side, we must solve the problem in terms of secondary quantities. It may be shown (see Problem 2.19) that the secondary currents of the transformer, corresponding to (2.31) and (2.32) (which are the primary currents), are

$$\mathbf{I}'_2 = \frac{-\mathbf{V}_t(a' - a'') + a''\mathbf{I}_2\mathbf{Z}''_{e2}}{a'\mathbf{Z}'_{e2} + a''\mathbf{Z}''_{e2}} \tag{2.33}$$

$$\mathbf{I}''_2 = \frac{\mathbf{V}_t(a' - a'') + a'\mathbf{I}_2\mathbf{Z}'_{e2}}{a'\mathbf{Z}'_{e2} + a''\mathbf{Z}''_{e2}} \tag{2.34}$$

where $\mathbf{Z}'_{e2}$ and $\mathbf{Z}''_{e2}$ are the equivalent impedances of the first and second transformers, respectively, referred to their secondaries.

We now substitute the following numerical values in (2.33) and (2.34):

$$a' = a'' = 10$$

$$\mathbf{Z}'_{e2} = \frac{1.84 \underline{/84.2}}{10^2} = 0.0184 \underline{/84.2} = (1.86 + j18.3) \times 10^{-3} \, \Omega$$

$$\mathbf{Z}''_{e2} = \frac{0.77 \underline{/82.5}}{10^2} = 0.0077 \underline{/82.5} = (1.0 + j7.63) \times 10^{-3} \, \Omega$$

$$I_2 = \frac{400{,}000}{230} = 1739 \text{ A}$$

and

$$V_t = 230 \text{ V}$$

Thus we obtain

$$\mathbf{I}_2' = \frac{10 \times 1739\underline{/-36.8} \times 0.0077\underline{/82.5}}{10(2.86 + j25.93) \times 10^{-3}} = 513.0\underline{/-38.1} \text{ A}$$

$$\mathbf{I}_2'' = \frac{10 \times 1739\underline{/-36.8} \times 0.0184\underline{/84.2}}{10(2.86 + j25.93) \times 10^{-3}} = 1226.0\underline{/-36.3} \text{ A}$$

The power supplied by the first transformer is

$$230 \times 513 \times \cos 38.1 = 92.7 \text{ kW}$$

The power supplied by the second transformer is

$$230 \times 1226 \times \cos 36.3 = 227.3 \text{ kW}$$

Check: The sum of the power supplied by the two transformers is

$$92.7 + 227.3 = 320 \text{ kW} = \text{total load} = 400 \times 0.8 = 320 \text{ kW}$$

From (2.33) and (2.34) it follows that load sharing by transformers in proportion to their ratings can be obtained by a proper choice of the respective $\mathbf{Z}_e$. ■

2.12 THREE-PHASE TRANSFORMER CONNECTIONS

Electric power systems are three-phase systems in that power is generated by three-phase generators and transmitted by three-phase transmission lines. Obviously, these generators and transmission lines must be linked by three-phase transformers. The primary and secondary windings of single-phase transformers may be interconnected to obtain three-phase transformer banks.

Some of the factors governing the choice of connections are as follows:

1. Availability of a neutral connection for grounding, protection, or load connections.
2. Insulation to ground and voltage stresses.
3. Availability of a path for the flow of third-harmonic (exciting) currents and zero-sequence (fault) currents.
4. Need for partial capacity with one unit out of service.
5. Parallel operation with other transformers.
6. Operation under fault conditions.
7. Economic considerations.

Keeping these factors in mind, we now consider some of the three-phase transformer connections.

Wye–Wye Connection. The wye–wye connection is shown in Fig. 2.26(a), where the terminal markings show subtractive polarities. Primary terminals are designated by *ABC*, whereas *abc* is used to indicate the secondary terminals. The phase relationships between the various voltages are shown in Fig. 2.26(b). The equilateral triangles superimposed on the phasor diagrams aid in the construction of the phasor diagrams. For instance *OA*, *OB*, and *OC* denote the phase voltages of the primary and $O'a$, $O'b$, and $O'c$ correspond to the phase voltages of the secondary. To determine the time–phase relationships of line voltages, we observe that the primary phase sequence is *ABC*. The phase of the voltage between *A* and *B* is found by tracing the circuit *AOB*, where *AO* is traversed in

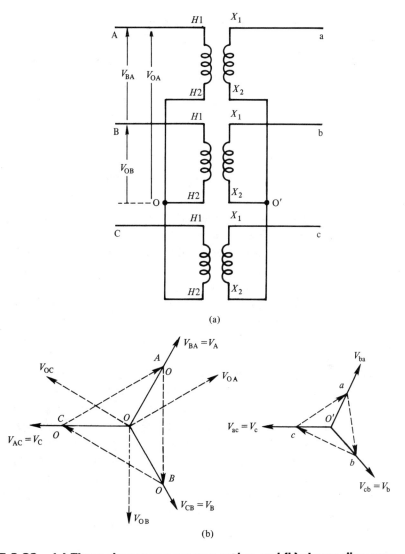

FIGURE 2-26. (a) Three-phase wye-wye connection, and (b) phasor diagrams.

2.12 THREE-PHASE TRANSFORMER CONNECTIONS

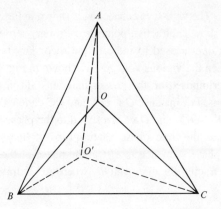

FIGURE 2-27. Roving neutral under unbalanced loads.

the positive direction and OB in the negative direction. The phasor relationship between the voltages around the circuit AOB is

$$\mathbf{V}_{BA} + \mathbf{V}_{OB} - \mathbf{V}_{OA} = 0$$

or

$$\mathbf{V}_{BA} = \mathbf{V}_{OA} - \mathbf{V}_{OB} = V_A$$

Thus we reverse OB and geometrically add it to OA to obtain BA, and hence $\mathbf{V}_{BA}$. Other voltages are determined in a similar fashion.

From Fig. 2.26(b) it is clear that there is no phase shift between the primary and secondary voltages. Furthermore, if V is the voltage between lines, then $V/\sqrt{3}$ is the voltage across the phase on the terminals of a wye-connected transformer. Thus compared to a delta connection (discussed later), the wye-connected transformer will have fewer turns, will require windings of larger cross section, and will have less dielectric stress on the insulation. On the other hand, the main disadvantage of wye–wye connected transformers is that such transformers have *roving neutrals* when supplying unbalanced loads. By "roving neutral" we mean that the potential of point O in Fig. 2.26(b) is not fixed with respect to the lines, and may take any position within the triangle if the transformer supplies an unbalanced load. Figure 2.27 shows two positions of O under two unbalanced loading conditions. One way to prevent the shifting of the neutral is to connect the primary neutral of the transformer to the neutral of the generator. In such a case, however, if there is a third harmonic component in the generator voltage waveform, there will be a third harmonic in the secondary voltage, and there will be corresponding triple frequency currents* in the secondary circuits.

*For a discussion of third-harmonic currents, see Section 2.18.

Delta–Delta Connection. Figure 2.28 shows three transformers connected to make a three-phase delta–delta system. The phasor diagram showing the phase relationships of various voltages is also included in Fig. 2.28. The transformers have subtractive polarities. Clearly, in such a connection, the individual windings of the transformers must be designed for full line voltages. This arrangement, however, has the advantage that one transformer can be out of service without an interruption in the system operation except that the capacity with two transformers is proportionately reduced. The secondary voltages also tend to be slightly unbalanced with the loss of one transformer. The delta–delta connection provides a path for the flow of the third-harmonic magnetizing current, the current in each coil being in phase with every other. Hence third-harmonic currents and voltages do not appear on the line.

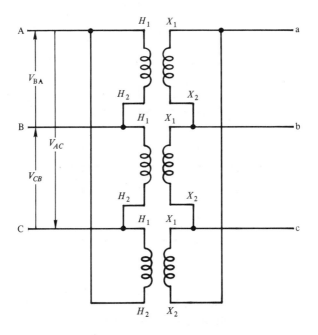

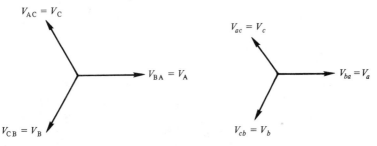

FIGURE 2-28. Delta-delta connection with subtractive polarities.

2.12 THREE-PHASE TRANSFORMER CONNECTIONS

Delta–Wye and Wye–Delta Connections. Delta–wye and wye–delta connections of three-phase transformers are shown in Figs. 2.29 and 2.30, respectively. The polarities in both cases are subtractive. Notice the 30° phase shift between the line and phase voltages in the two cases. This shift in the delta–wye connection is opposite that in the wye–delta connection. These connections are particularly suited for high-voltage systems. The delta–wye connection is used for stepping up and the wye–delta for stepping down the voltage. The wye connection on the high-voltage side permits the grounding of the neutral. The delta connection offers a path for the flow of the third-harmonic currents.

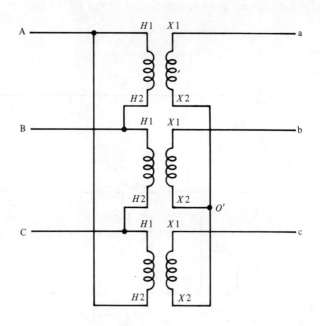

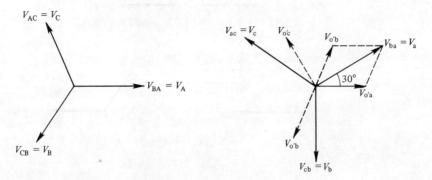

FIGURE 2-29. Delta-wye connection.

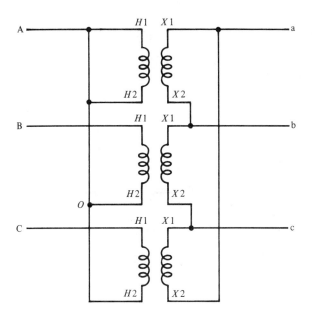

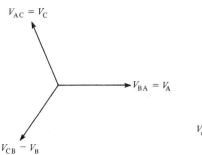
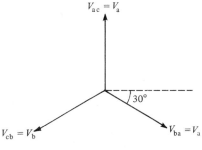

FIGURE 2-30. Wye-delta connection.

2.13

CERTAIN SPECIAL TRANSFORMER CONNECTIONS

In the preceding section we studied three-phase transformer connections, where the primaries and/or the secondaries are connected in wye and/or delta. Such connections involved three transformers. There are certain three-phase transformer connections which require only two transformers. We discuss three such connections here.

Open-Delta Connection. Recall from the preceding section that in the delta–delta connection if one of the transformers is removed, the remaining two transformers can provide a three-phase system. In practice, such a connection

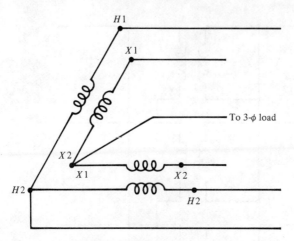

FIGURE 2-31. Open-delta Connection.

is known as the *open-delta* or *V connection,* and is shown in Fig. 2.31. Obviously, the rating of an open-delta transformer bank is less than that of a delta–delta bank (assuming that each transformer of the two banks has the same rating). If each transformer of a delta–delta system is rated at V volts and I amperes, the line current for the system will be $\sqrt{3}\ I$. But for an open-delta system, the line current cannot exceed I. Consequently, the combined rating of two transformers connected in open-delta system amounts to $I/\sqrt{3}\ I$, or 58 percent of the rating of the original three transformers.

Tee Connection. The tee-connection also offers a method of using only two transformers in a three-phase system. This connection is shown in Fig. 2.32(a). One transformer, having a center tap at O, is connected to the terminals A and B of the three-phase system. This transformer is called the *main transformer.* The second transformer, called the *teaser transformer,* has one terminal connected to point O of the main transformer and the other to C of the three-

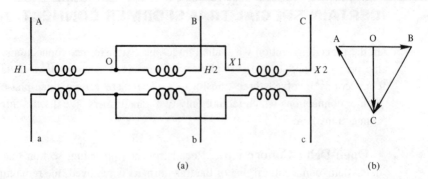

FIGURE 2-32. (a) T-connection; (b) phasor diagram.

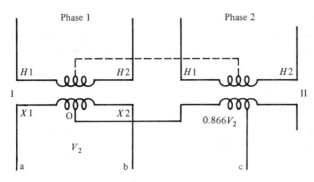

FIGURE 2-33. Scott connection.

phase source, as indicated in Fig. 2.32(a). From the phasor diagram in Fig. 2.32(b), it may be seen that the voltage V_{OC} across the teaser transformer = $(\sqrt{3}/2)V_{AB} = 0.866V_{AB}$. The ratio of the transformation of the two transformers being the same, we obtain a three-phase balanced system at the secondary. It is important that relative polarities be properly maintained, as shown in Fig. 2.32(a). With the correct polarities, the currents in each of the halves of the main transformer flow in opposite directions.

Scott Connection. The Scott connection is a special connection used to obtain a two-phase system from a three-phase system, and vice versa. The connection is shown in Fig. 2.33. Transformer I has a midpoint tap and has a voltage rating V_2. Transformer II has a tap at a point corresponding to 86.6 percent of V_2.

There are numerous other transformer connections for various applications, but these are not considered here.

2.14 PARALLEL OPERATION OF THREE-PHASE TRANSFORMERS

In operating three-phase transformers in parallel, it is most important that there be no phase displacement between the voltages at the secondary leads which are being connected in parallel. To verify if this condition is fulfilled, we may refer to the phasor diagrams of three-phase transformer connections. For instance, the phasor diagrams of Figs. 2.26 and 2.28 indicate that a wye–wye connected transformer bank will operate in parallel with a delta–delta connected bank if the line-to-line voltages of the two banks are the same. Notice from the respective phasor diagrams that the two phasor diagrams are identical. On the other hand, a wye–wye or a delta–delta bank will not operate satisfactorily in parallel with either a wye–delta or a delta–wye bank. In such connections, when the two sets of primary leads are connected in parallel and to the same source, there will always be a phase displacement between the two sets of secondary leads.

2.15 AUTOTRANSFORMERS

In contrast to the two-winding transformers considered so far, the autotransformer is a single-winding transformer having a tap brought out at an intermediate point. Thus, as shown in Fig. 2.34, ac is the single winding (wound on a laminated core) and b is the intermediate point where the tap is brought out. The autotransformer may be used as either a step-up or a step-down operation, like a two-winding transformer. Considering a step-down arrangement, let the primary applied (terminal) voltage be V_1, resulting in a magnetizing current and a core flux, φ_m. Let the secondary be open-circuited. Then the primary and secondary voltages obey the same rules as in a two-winding transformer, and we have

$$\frac{V_1}{V_2} = \frac{E_1}{E_2} = \frac{N_1}{N_2} = a$$

with $a > 1$ for step-down.

Furthermore, ideally,

$$V_1 I_1 = V_2 I_2$$

and

$$\frac{V_1}{V_2} = \frac{I_2}{I_1} = a$$

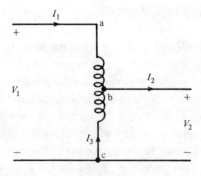

FIGURE 2-34. A step-down autotransformer.

Neglecting the magnetizing current, we must have the mmf balance equation as

$$N_2 I_3 = (N_1 - N_2) I_1$$

or

$$I_3 = \frac{N_1 - N_2}{N_2} I_1 = (a - 1) I_1 = I_2 - I_1$$

which agrees with the current-flow directions shown in Fig. 2.34.

The apparent power delivered to the load may be written as

$$P = V_2 I_2 = V_2 I_1 + V_2 (I_2 - I_1) \qquad (2.35)$$

In (2.35) the power is considered to consist of two parts:

$$V_2 I_1 \equiv P_c \equiv \text{conductively transferred power through } bc$$

$$V_2 (I_2 - I_1) \equiv P_i \equiv \text{inductively transferred power through } ab$$

These powers are related to the total power by

$$\frac{P_i}{P} = \frac{I_2 - I_1}{I_2} = \frac{a - 1}{a} \qquad (2.36)$$

and

$$\frac{P_c}{P} = \frac{I_1}{I_2} = \frac{1}{a} \qquad (2.37)$$

where $a > 1$.

It may be shown that for a step-up transformer the power ratios are obtained as follows:

$$P = V_1 I_1 = V_1 I_2 + V_1 (I_1 - I_2)$$

implying that the total apparent power consists of two parts:

$$P_c = V_1 I_2 = \text{conductively transferred power}$$

$$P_i = V_1 (I_1 - I_2) = \text{inductively transferred power}$$

Hence

2.15 AUTOTRANSFORMERS

$$\frac{P_i}{P} = \frac{I_1 - I_2}{I_1} = 1 - a \tag{2.38}$$

and

$$\frac{P_c}{P} = \frac{I_2}{I_1} = a \tag{2.39}$$

where $a < 1$.

EXAMPLE 2.15

A 10-kVA 440/110-V two-winding transformer is reconnected as a step-down 550/440-V autotransformer. Compare the volt-ampere rating of the autotransformer with that of the original two-winding transformer, and calculate P_i and P_c.

SOLUTION

Refer to Fig. 2.34. The rated current in the 110-V winding (or in ab) is

$$I_1 = \frac{10{,}000}{110} = 90.91 \text{ A}$$

The current in the 440-V winding (or in bc) is

$$I_3 = I_2 - I_1 = \frac{10{,}000}{440} = 22.73 \text{ A}$$

which is the rated current of the winding bc. Thus the load current is

$$I_2 = I_1 + I_3 = 90.91 + 22.73 = 113.64 \text{ A}$$

Check: For the autotransformer

$$a = \frac{550}{440} = 1.25$$

and

$$I_2 = aI_1 = 1.25 \times \frac{10{,}000}{110} = 113.64 \text{ A}$$

which agrees with I_2 calculated above. Hence the rating of the autotransformer is

$$P_{auto} = V_1 I_1 = 550 \times \frac{10,000}{110} = 50 \text{ kVA}$$

Thus the inductively supplied apparent power is

$$P_i = V_2(I_2 - I_1) = \frac{a-1}{a} P = \frac{1.25 - 1}{1.25} \times 50 = 10 \text{ kVa}$$

which is the volt-ampere rating of the two-winding transformer.
The conductively supplied power is

$$P_c = \frac{P}{a} = \frac{50}{1.25} = 40 \text{ kVA}$$

■

EXAMPLE 2.16
Repeat the problem of Example 2.15 for a 440/550-V step-up connection.

SOLUTION
The step-up connection is shown in Fig. 2.35. The rating of the winding *ab* is 110 V and the load current I_2 flows through *ab*. Hence

$$I_2 = \frac{10,000}{110} = 90.91 \text{ A}$$

The output voltage is

$$V_2 = 550 \text{ V}$$

Thus the volt-ampere rating of the autotransformer is

$$V_2 I_2 = 550 \times \frac{10,000}{110} = 50 \text{ kVA}$$

which is the same as in Example 2.15.

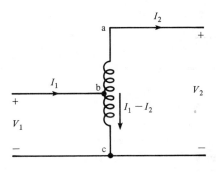

FIGURE 2-35. A step-up autotransformer.

2.15 AUTOTRANSFORMERS

Thus the power transferred conductively is

$$V_1 I_2 = 440 \times 90.91 = 40 \text{ kVA}$$

and the power transferred inductively is

$$50 - 40 = 10 \text{ kVA}$$

Consequently, a two-winding transformer connected as an autotransformer will have a volt-ampere rating $a/(a - 1)$ times its rating as a two-winding transformer.

2.16

THREE-WINDING TRANSFORMERS

Generally, large power transformers have three windings. The third winding is known as a *tertiary winding,* which may be used for the following purposes:

1. To supply a load at a voltage different from the secondary voltage.
2. To provide a low impedance for the flow of certain abnormal currents, such as third-harmonic currents.
3. To provide for the excitation of a regulating transformer.

Wherever more than one secondary voltage is required, it can be secured more economically from a third winding than from an additional transformer.

A three-winding transformer is represented schematically in Fig. 2.36. When a three-winding transformer operates on load, each winding interacts inductively with each of the others. Thus there are three mutual inductances and three self- (or leakage) inductances to be considered. The analysis therefore becomes rather cumbersome and is not considered here.

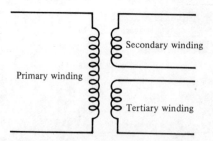

FIGURE 2-36. A three-winding transformer.

2.17 INSTRUMENT TRANSFORMERS

Instrument transformers are of two kinds: current transformers (CTs) and potential transformers (PTs). These are used to supply power to ammeters, voltmeters, wattmeters, relays, and so on. Instrument transformers are used for (1) reducing the measured quantity to a low value which can be indicated by standard instruments (a standard voltmeter may be rated at 120 V and an ammeter at 5 A), and (2) isolating the instruments from high-voltage sources for safety. A connection diagram of a CT and a PT with an ammeter, a voltmeter, and a wattmeter is shown in Fig. 2.37. The load on the instrument transformer is called the *burden*. Depending on the burden, instrument transformers are rated from 25 to 500 VA. However, a PT or a CT is much (two to six times) bigger than a power transformer of the same rating.

An ideal instrument transformer has no phase difference between the primary and secondary voltages (or currents), which are independent of the burden. Like the ideal power transformer discussed earlier, the voltage ratio of an ideal PT is exactly equal to its turns ratio. The current ratio of an ideal CT is exactly equal to the inverse of the turns ratio. In practice, however, load-dependent ratio and phase-angle errors are present in instrument transformers.

The principle of operation of an instrument transformer is no different from that of an ordinary power transformer. Thus they have similar equivalent circuits and phasor diagrams, as shown in Fig. 2.38. It is clear from this diagram that the secondary impedance drop causes a phase displacement α, and the primary impedance drop a phase displacement β; the exciting current I_0 causes a further phase displacement γ, so that the angle between the primary voltage and current is $(\theta_2 + \alpha + \beta + \gamma)$, compared with an angle θ_2 between the secondary voltage and current. Thus the transformer introduces a phase-angle error $(\alpha + \beta + \gamma)$. Moreover, V_1 and V_2 will be only approximately in the ratio of

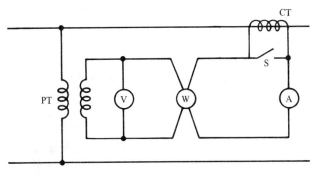

FIGURE 2-37. Instrument transformer connections.

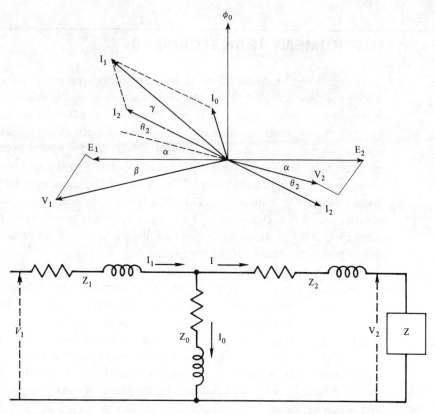

FIGURE 2-38. Phasor diagram and equivalent circuit of an instrument transformer.

the number of turns. The significance of these errors is indicated in Fig. 2.39. Calibration curves of this type are furnished by the manufacturer. To nullify or reduce the errors, instrument transformers are designed with (1) small leakage reactances and low resistances that reduce angles α and β; (2) low flux densities and good transformer iron, which reduce the exciting current I_0 and therefore angle γ; and (3) a less-than-nominal turn ratio, which compensates for the ratio error. Compensating impedances may also be provided, so that the burden can be kept constant as instruments are put in or taken out of the circuit. For a constant burden, the instruments may be calibrated, or corrected, against the load.

Provision must be made to short-circuit the secondary of a current transformer before removing any instruments, for otherwise dangerously high voltages may occur. It is clear from the equivalent circuit of Fig. 2.38 that if the secondary is open-circuited, the primary current, of fixed magnitude as determined by the load, must flow through the exciting impedance and act entirely as a magnetizing current to the transformer. This results in high flux density and correspondingly high voltage. Oscillograms of such a voltage show it to be peaked (see the next section) by a dominating third harmonic, as would be expected.

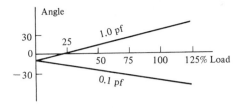
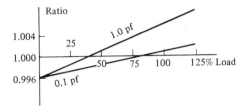

FIGURE 2-39. Phase angle and ratio error of an instrument transformer.

2.18 THIRD HARMONICS IN TRANSFORMERS

In earlier sections we have mentioned the presence of third-harmonic currents in transformers. First, we show a sine wave (or the fundamental) and its third harmonic in Fig. 2.40. The resultant, which is the sum of the fundamental and the third harmonic, is also shown in Fig. 2.40. Next, let us assume that the core flux of a transformer is sinusoidal. The corresponding core flux density, $B(t)$, is shown in Fig. 2.41, which also shows the core saturation (BH) characteristic. From these $B(t)$ and BH curves we obtain, graphically, the waveform of the

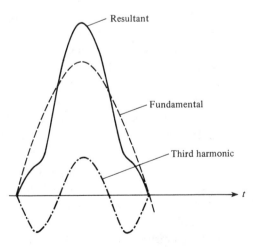

FIGURE 2-40. Resultant waveform of a fundamental and third harmonic.

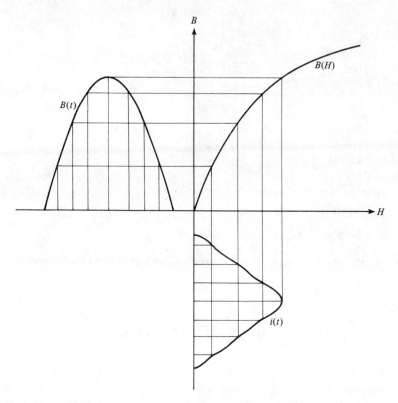

FIGURE 2-41. Effect of saturation on exciting current.

corresponding magnetizing current. The construction procedure and the current waveform are given in Fig. 2.41. Now, compare the resultant waveform of Fig. 2.40 and the magnetizing current waveform of Fig. 2.41. The similarity between the two indicates that the magnetizing current does, indeed, contain third harmonics if the transformer core operates under a saturated condition.

Having established the presence of the third harmonic in a transformer, let us now consider their mutual relationships in a three-phase transformer. In a three-phase system, the third harmonic of the phase displaced from the reference phase by 120° is displaced by 3 × 120° = 360°. Also, the third harmonic in the phase displaced from the reference phase by 240° is displaced by 3 × 240° = 720°. Consequently, in a three-phase transformer, the third harmonics in the three transformers are in phase. In a delta connection, third harmonics of a current add to each other and form circulating currents around the delta. This connection is, therefore, called a *trap* for third harmonics. In an ungrounded wye connection, the third harmonics of voltage add to each other and tend to push the neutral off the geometric neutral (see Fig. 2.27).

QUESTIONS

2.1 List some applications for transformers.

2.2 What is the difference between a power transformer and a transformer used in an electrical communication circuit?

2.3 What is the difference between a power transformer and an instrument transformer?

2.4 How is an autotransformer different from a two-winding transformer?

2.5 What features distinguish an ideal transformer from an actual transformer?

2.6 What is the principle of operation of a transformer? Why will a transformer not function normally when supplied from a dc source?

2.7 How are the polarities of a two-winding transformer identified?

2.8 In a step-up transformer is (a) the primary voltage greater or less than the secondary voltage? (b) the primary current greater or less than the secondary current?

2.9 Do the leakage reactances have an effect on (a) the efficiency of a transformer? (b) the voltage regulation of the transformer?

2.10 Do the winding resistances have an effect on (a) the efficiency and (b) the voltage regulation of the transformer?

2.11 Is it desirable to have a large or small (a) leakage reactance and (b) magnetizing reactance in a power transformer? Why?

2.12 Is all-day efficiency a measure of the power efficiency or energy efficiency of a transformer?

2.13 What are major losses present in a transformer? How are these losses minimized?

2.14 What approximations are made to reduce the exact equivalent circuit to an approximate equivalent circuit of a transformer?

2.15 Name the tests that may be performed to obtain the approximate equivalent circuit of a transformer.

2.16 Name the parameters of the approximate equivalent circuit that account for (a) leakages, (b) core losses, (c) winding losses, and (d) magnetization of a transformer.

2.17 For the same power rating, will a 60-Hz transformer be heavier or lighter than a 400-Hz transformer?

2.18 A transformer is excited by a sinusoidal voltage source. It is found that the current is nonsinusoidal. Why?

2.19 For what purpose is the Scott connection of transformers used?

2.20 Will a wye–wye transformer bank operate satisfactorily in parallel with (a) a delta–delta bank? (b) a wye–delta bank?

2.21 What are the advantages of autotransformers over two-winding transformers?

2.22 What are the applications of instrument transformers?

PROBLEMS

2.1 Defining the form factor of a waveform by (2.10), show that the form factor of a sine wave is 1.11.

2.2 A 220/110-V 60-Hz ideal transformer is rated at 5 kVA. Calculate (a) the turns ratio and (b) the primary and secondary currents (1) at full load and (2) at a 3-kW load at 0.8 lagging.

2.3 A 10-Ω resistance is connected across the secondary of a 220/110-V 60-Hz ideal transformer. The secondary voltage is 110 V. Calculate (a) the primary current and (b) the equivalent resistance referred to the primary. (c) Determine the power dissipated in this equivalent resistance if the primary current calculated in part (a) flows (through this resistance).

2.4 The secondary winding of a 60-Hz transformer has 50 turns. If the induced voltage in this winding is 220 V, determine the maximum value of the core flux.

2.5 A 60-Hz transformer having a 300-turn primary takes 60 W of power and 0.8 A of current at an input voltage of 110 V. The resistance of the winding is

0.2 Ω. Calculate (a) the core loss; (b) the magnetizing reactance, X_m; and (c) the core loss equivalent resistance, R_c. Neglect leakage impedance drop.

2.6 Repeat the calculations of Example 2.6 using the circuit referred to the secondary. Neglect the primary leakage impedance drop.

2.7 Refer to the data of the transformer of Example 2.6. Using the equivalent circuit referred to the secondary, calculate the efficiency of the transformer at full load and 0.8 leading power factor.

2.8 Verify that the circuit shown in Fig. 2.13 is a valid equivalent circuit of a transformer referred to its secondary side.

2.9 Two 1150/115-V transformers are given. Show the interconnections of these transformers for (a) 2300/115-V operation and (b) 1150/115-V operation. Indicate the polarity markings.

2.10 A 25-kVA 440/220-V 60-Hz transformer has a constant core loss of 710 W. The equivalent resistance and reactance referred to the primary are 0.16 Ω and 0.66 Ω, respectively. Plot the following curves: (a) core loss versus primary current, (b) I^2R loss versus primary current, and (c) efficiency versus primary current. The primary current to range from 0 to 100 A. What is the I^2R loss when the efficiency of the transformer is maximum? (*Note:* This is a special case of the general rule that the efficiency of a transformer is maximum when the I^2R losses in the two windings equal the core losses.)

2.11 A 75-kVA transformer has an iron loss of 1 kW and a full-load copper loss of 1 kW. Calculate the transformer efficiency at unity power factor at (a) full-load and (b) half full-load.

2.12 During one day, the transformer of Problem 2.11 operates on full-load at unity power factor for 8 h, on no-load for 8 h, and on one-half load at unity power factor for 8 h. Determine the all-day efficiency.

2.13 The maximum efficiency of a 100-kVA transformer is 98.4 percent and occurs at 90 percent full-load at unity power factor. Calculate the efficiency of the transformer at unity power factor at (a) full-load and (b) 50 percent full-load.

2.14 The results of open-circuit and short-circuit tests on a 10-kVA 440/220-V 60-Hz transformer are as follows:

Open-circuit test: high-voltage side open: $V_0 = 220$ V, $I_0 = 1.2$ A, and $W_0 = 150$ W

Short-circuit test: low-voltage side short-circuited: $V_s = 20.5$ V, $I_s = 42$ A, and $W_s = 140$ W

Determine the parameters of the approximate equivalent circuit from the data given. Refer the circuit to HV side.

2.15 Calculate the primary voltage of the transformer of Problem 2.14 when the secondary is supplying full-load at 0.8 lagging power factor. What is the efficiency of the transformer at this load?

2.16 A 1000-kVA transformer has a 94 percent efficiency at full-load and at 50 percent full-load. The power factor is unity in both cases. (a) Segregate the losses. (b) For unity power factor and 75 percent full-load, determine the efficiency of the transformer.

2.17 The primary and secondary voltages of an autotransformer are 440 V and 360 V, respectively. If the secondary current is 80 A, determine the primary current.

2.18 A 5-kVA 60-Hz two-winding transformer has a full-load efficiency of 96 percent at unity power factor. The iron loss at 60 Hz is 60 W. This transformer is next connected as an autotransformer, and delivers a 5-kVA unity power factor load at 220 V. The primary of the autotransformer is connected to a 440-V source. Calculate (a) the primary current and (b) the efficiency of the autotransformer.

2.19 A 100-kVA 42,000/2400-V 60-Hz transformer is operated in parallel with a 75-kVA 42,000/2400-V 60-Hz transformer. The respective impedances referred to the secondary are $(0.5 + j4)$ Ω and $(0.8 + j6)$ Ω. The total load on the transformer is 120 kVA at 0.8 lagging power factor. Calculate the load on each transformer.

2.20 Calculate the primary and the secondary currents for the two transformers of Problem 2.19.

2.21 Three single-phase transformers are connected to make a three-phase delta–wye bank. The line voltages of the bank are 2200 V and 400 V on the primary and secondary, respectively. The load on the transformer is 15 kW, 0.8 power factor lagging, wye connected. Determine the primary and secondary line and phase currents.

2.22 A 100-kVA 2200/220-V wye–wye three-phase 60-Hz transformer bank has an iron loss of 1500 W and a copper loss of 2700 W at full-load. Determine the efficiency of the transformer for 75 percent full-load at 0.8 power factor.

2.23 Two transformers, each rated at 50 kVA, are connected in open-delta to form a three-phase system. What total kVA can be obtained from the open-delta bank without overloading the transformers?

2.24 Draw the voltage phasor diagrams of wye–wye- and delta–wye-connected transformers. Verify that the two sets of secondary voltages are 30° apart. Under these conditions, should these two banks be operated in parallel?

2.25 A three-phase 75-kVA 440-V 0.8 lagging power factor load is to be supplied by single-phase transformers having a turns ratio of 2. Calculate the currents in the windings of each transformer and the corresponding power factor if the transformers are connected in (a) open-delta and (b) delta–delta.

2.26 Two Scott-connected transformers are supplied from a 400-V three-phase source. The two-phase secondary supplies a 100-kVA load at 220 V. Determine the currents in the windings of the two transformers.

2.27 A current transformer (CT) and a potential transformer (PT) are connected to the current and voltage coils, respectively, of a wattmeter measuring power going into a load. The wattmeter reading is 240 W, and the turns ratios of the CT and the PT are 100:5 and 10:1, respectively. What is the actual power supplied to the load?

CHAPTER 3

DC Machines

3.1 INTRODUCTION

In Chapter 2 we studied the transformer. In a sense the transformer is an *energy transfer* device—energy is transferred from the primary to the secondary. During the energy transfer process, the form of energy remains unchanged; that is, the energy at both the input (primary) and output (secondary) terminals is electrical. In contrast, in a rotating electric machine, electrical energy is converted into mechanical form, and vice versa, depending on the mode of operation of the machine. An *electric motor* converts electrical energy into mechanical energy, whereas an *electric generator* is used to convert mechanical energy into electrical energy. For this reason, electric machines are also called *electromechanical energy converters*. In essence, the process of electromechanical energy conversion can be expressed as

$$\text{electrical energy} \underset{\text{generator}}{\overset{\text{motor}}{\rightleftharpoons}} \text{mechanical energy}$$

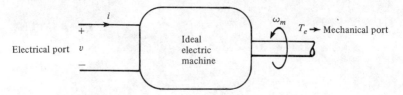

FIGURE 3-1. A general representation of an electric machine.

Schematically, an ideal electric machine may be represented by Fig. 3.1, for which we have, over a certain time interval Δt,

$$(vi)\,\Delta t = (T_e \omega_m)\,\Delta t$$

or

$$vi = T_e \omega_m \tag{3.1}$$

where v and i are, respectively, the voltage and current at the electrical port, and T_e and ω_m are, respectively, the torque (in newton-meters, N-m) and angular rotational velocity (in radians per second, rad/s) at the mechanical port. We wish to reiterate that (3.1) is valid for an ideal machine in that the machine is lossless. (We will have more to say later about losses in machines.)

The remainder of the book will be concerned with electric machines, beginning with dc machines. The dc machine is the first machine devised for electromechanical energy conversion. The simplest form of the dc machine is the Faraday disk, which is discussed in the next section.

3.2

THE FARADAY DISK AND FARADAY'S LAW

Based on the principle of electromagnetic induction, in 1832 Faraday demonstrated that if a copper disk is rotated in an axially directed magnetic field (produced by a permanent magnet), with sliding contacts, or brushes, mounted at the rim and at the center of the disk, a voltage will be available at the brushes. Such a machine is shown in Fig. 3.2 and is commonly known as the Faraday disk or *homopolar* generator. Another name for the homopolar machine is the *acyclic* machine. The device shown in Fig. 3.2 is also capable of operating as a motor. It is a homopolar machine because the conductors (which may be imagined in form of spokes of the disk) are under the influence of the magnetic field of one polarity at all times. In contrast, in *heteropolar* machines the moving conductors are under the influence of magnetic fields of opposite polarities in an alternating fashion, as we shall soon see.

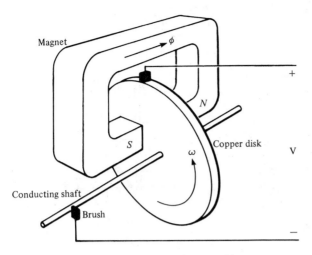

FIGURE 3-2. Faraday disk or homopolar machine.

As mentioned earlier, the operation of the homopolar generator is based on Faraday's law of electromagnetic induction. The law states that an emf is induced in a circuit placed in a magnetic field if either (1) the magnetic flux linking the circuit is time varying, or (2) there is a relative motion between the circuit and the magnetic field such that the conductors comprising the circuit cut across the magnetic flux lines. The first form of the law, stated as (1), is the basis of operation of transformers (see Sections 1.7 and 2.4). The second form, stated as (2), is the basic principle of operation of electric generators, including the homopolar generator.

Consider a conductor of length l located at right angles to a uniform magnetic field of flux density B as shown in Fig. 3.3. Let the conductor be connected with fixed external connections to form a closed circuit. These external connections are shown in the form of conducting rails in Fig. 3.3. The conductor can slide on the rails, and thereby the flux linking the circuit changes. Hence, according to Faraday's law, an emf will be induced in the circuit, and a voltage, v, will be measured by the voltmeter. If the conductor moves with a velocity u (m/s) in a direction at right angles to B and l both, the area swept by the conductor in 1 second is lu. The flux in this area is Blu, which is also the flux linkage (since in effect, we have a single-turn coil formed by the conductor, the rails, and the voltmeter). In other words, flux linkage per unit time is Blu, which is, then, the induced emf, e. We write this in the equation form as

$$e = Blu \qquad (3.2)$$

This form of Faraday's law is also known as the *flux-cutting rule*. Stated in words, an emf, e, as given by (3.2), is induced in a conductor of length l if it "cuts" magnetic flux lines of density B by moving at right angles to B at a

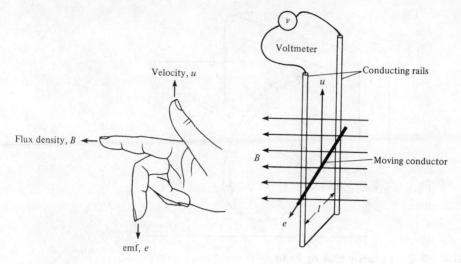

FIGURE 3-3. The right-hand rule.

velocity u (at right angles to l). The mutual relationships among e, B, l, and u are given by the *right-hand rule*, as shown in Fig. 3.3. Compare with the left-hand rule of Fig. 1.2.

EXAMPLE 3.1
A conducting disk of 0.5 m radius rotates at 1200 rpm in a uniform magnetic field of 0.4 T. Calculate the voltage available between the rim and the center of the disk.

SOLUTION
Referring to (3.2), in this problem we have $B = 0.4$ T, $l = 0.5$ m, and $u = r\omega$, where $\omega = 2\pi n/60$ rad/s is the angular velocity of the disk. Substituting $n = 1200$ rpm and $r = 0.5$ m, we obtain the linear velocity of the rim as

$$u_{\text{rim}} = 0.5 \times 2\pi \times \frac{1200}{60} = 20\pi \text{ m/s}$$

But the linear velocity of the center of the disk is zero. Hence the average velocity of the disk (or a radial conductor) is given by

$$u = \tfrac{1}{2}(20\pi + 0) = 10\pi \text{ m/s}$$

and, from (3.2),

$$e = 0.4 \times 0.5 \times 10\pi = 6.28 \text{ V}$$

Let us now summarize the salient points of this section: (1) The Faraday disk operates on the principle of electromagnetic induction, enunciated by Faraday. (2) There are two forms of expression of Faraday's law: time rate of change of flux linkage (with a circuit), or flux cut (by a conductor) results in an induced emf, but the latter is contained in the former. (3) Electric generators operate on the flux cutting principle. (4) Example 3.1 shows that rather a small voltage (12.76 V) is induced in a disk of large diameter (1 m) rotating at a reasonably high speed (1200 rpm). Thus the Faraday disk in its primitive form is not a practical type of dc generator. Indeed, present-day dc generators have little resemblance to the Faraday disk. However, the primitive Faraday disk generator has been developed into present-day homopolar, or acyclic generators which find applications requiring very large currents at very low voltages. A typical example is that of the chlorine-cell line in the chemical industry. Such electrochemical loads require a few thousand amperes of current per volt. Another example of application of the homopolar generator is in an aluminum-pot line requiring a current of over 150,000 A at 400 V.

3.3

THE HETEROPOLAR OR CONVENTIONAL DC MACHINE

Consider an N-turn coil, rotating at a constant angular velocity ω in a uniform magnetic field of flux density B, as shown in Fig. 3.4(a). Let l be the axial length of the coil and r be its radius. The emf enduced in the coil can be found by an application of (3.2). However, we must be careful in determining u in (3.2). Recall from the preceding section that u is the velocity at right angles to B. In the system under consideration, we have a rotating coil. Thus u in (3.2) corresponds to that component of velocity which is at right angles to B. We illustrate the components of velocities in Fig. 3.4(b), where the tangential velocity $u_t = r\omega$ has been resolved into a component u_1, along the direction of the flux, and another component u_2, across (or perpendicular to) the flux. The latter component "cuts" the magnetic flux, and is the component responsible for the emf induced in the coil. The u in (3.2) should then be replaced by u_2. The next term in (3.2) that needs careful consideration is l_1, the effective length of the conductor. For the N-turn coil of Fig. 3.4(a), we effectively have $2N$ conductors in series (since each coil side has N conductors and there are two coil sides). If l_1 is the length of each conductor, then the total effective length of the N-turn coil is $2Nl_1$, which should be substituted for l in (3.2). The form of (3.2) for the N-turn coil then becomes

$$e = B(2Nl_1)u_2 \qquad (3.3)$$

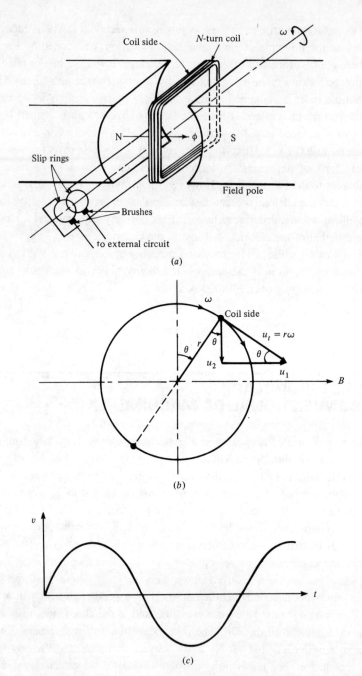

FIGURE 3-4. (a) An elementary generator; (b) resolution of velocities into "parallel" and "perpendicular" components; (c) voltage wave form at the brushes.

From Fig. 3.4(b), we have

$$u_2 = u_t \sin\theta = r\omega \sin\theta \tag{3.4}$$

If the coil rotates at a constant angular velocity ω, then

$$\theta = \omega t \tag{3.5}$$

Consequently, (3.3) to (3.5) yield

$$e = 2BNl_1 r\omega \sin\omega t = E_m \sin\omega t \tag{3.6}$$

where $E_m = 2BNl_1 r\omega$. A plot of (3.6) is shown in Fig. 3.4(c). The conclusion is that a sinusoidally varying voltage will be available at the slip rings, or brushes, of the rotating coil shown in Fig. 3.4(a). The brushes reverse polarities periodically.

In order to obtain a unidirectional voltage at the coil terminals, we replace the slip rings of Fig. 3.4(a) by the *commutator segments* shown in Fig. 3.5(a). It can be readily verified, by applying the right-hand rule, that in the arrangement shown in Fig. 3.5(a) the brushes will maintain their polarities regardless of the position of the coil. In other words, brush *a* will always be positive, and brush *b* will always be negative for the given relative polarities of the flux and the direction of rotation. Thus the arrangement shown in Fig. 3.5 forms an elementary *heteropolar* dc machine. The main characteristic of a heteropolar dc machine is that the emf induced in a conductor, or a current flowing through it, has its direction reversed as it passes from a north-pole to a south-pole region. This reversal process is known as *commutation,* and is accomplished by the commutator-brush mechanism, which also serves as a connection to the external circuit. The voltage available at the brushes will be of the form shown in Fig. 3.5(b). In this respect, commutation is the process of rectification of the induced alternating emf in the coil into a dc voltage at the terminals.

The natural question to ask at this point is: What have we gained, compared to the Faraday disk, by introducing the complications of a commutator? The answer to this question lies in the fact that a full range of ratings can be realized from a heteropolar cylindrical configuration. Recall from Example 3.1 that we could get only 13 V at the terminals of a homopolar machine. Obtaining much higher voltages is no problem with a heteropolar machine, as we shall presently see. As a matter of convention, we will term the dc heteropolar commutator machine simply a dc machine (unless otherwise stated).

Before leaving the subject of the homopolar and the elementary heteropolar machines, let us recall (3.1), according to which these machines must also be able to operate (at least in principle) as motors. The operation of the homopolar machine as a dc motor is left as an exercise for the reader. For the heteropolar machine, the production of a unidirectional torque, and hence operation as a dc

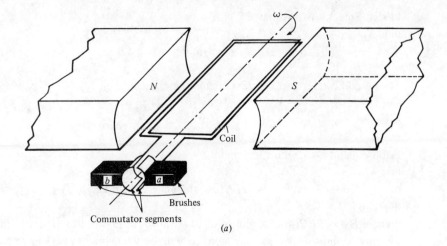

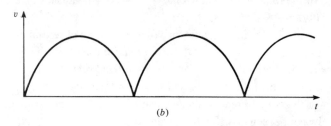

FIGURE 3-5. (a) An elementary dc generator; (b) output voltage at the brushes.

motor can be verified by referring to Fig. 3.6. Conductors a and b are the two sides of a coil. Thus a and b are connected in series. Notice from Fig. 3.6(a) that the current enters into conductor a through brush a and leaves conductor b through brush b. Applying the left-hand rule, the conductors will experience a force to produce a counterclockwise rotation. After half a rotation, the conductors interchange their respective position, as shown in Fig. 3.6(b). Now the current in conductor b is again in such a direction that the force on the coil will

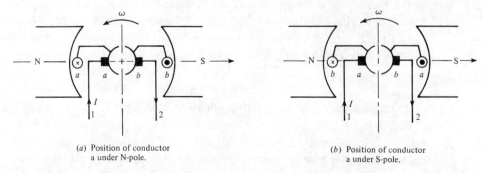

FIGURE 3-6. Production of a unidirectional torque and operation of an elementary dc motor.

tend to rotate it in a counterclockwise direction. In other words, the torque is unidirectional and is independent of conductor position.

3.4 CONSTRUCTIONAL DETAILS

Before we consider some conventional dc machines, it is best that we study briefly the constructional features of their various parts and discuss the usefulness of these parts. We observe from the preceding section that the basic elements of a dc machine are the rotating coil, a means for the production of flux, and the commutator-brush arrangement. In a practical dc machine the coil is replaced by the *armature winding* mounted on a cylindrical magnetic structure. The flux is provided by the *field winding* wound on field poles. Generally, the armature winding is placed on the rotating member—the *rotor*—and the field winding is on the stationary member—the *stator* of the dc machine. A common large dc machine is shown in Fig. 3.7, which, together with the schematic of Fig. 3.8, shows most of the important parts of a dc machine. The *field poles*, mounted on the stator, carry the field windings. Some machines carry more than one independent field winding on the same core. The cores of the poles are built of sheet-steel laminations. Because the field windings carry direct current, it is not

FIGURE 3-7. A dc machine (Courtesy of General Electric Company).

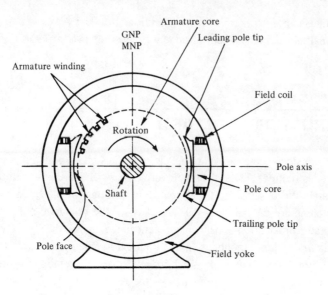

FIGURE 3-8. Parts of a dc machine.

necessary to have the pole cores laminated. It is, however, necessary for the pole faces to be laminated because of their proximity to the armature windings. (Use of laminations for the cores as well as for the pole faces facilitates assembly.) The rotor or the armature core, which carries the rotor or armature windings, is generally made of sheet-steel laminations. These laminations are stacked together to form a cylindrical structure. On its outer periphery the armature (or rotor) has *slots* in which the armature coils that make up the armature winding are located. For mechanical support, protection from abrasion, and for greater electrical insulation, nonconducting slot liners are often wedged in between the coils and the slot walls. The magnetic material between the slots comprises the *teeth*. A typical slot/tooth geometry for a large dc machine is shown in Fig. 3.9. The commutator is made of hard-drawn copper segments insulated from one another by mica. The details of the commutator assembly are given in Fig. 3.10. The armature windings are connected to the commutator segments over which the carbon brushes slide and serve as leads for electrical connection.

The armature winding may be a *lap winding* [Fig. 3.11(a)] or a wave winding [Fig. 3.11(b)], and the various coils forming the armature winding may be connected in a series–parallel combination. In practice, the armature winding is housed as two layers in the slots of the armature core. In large machines the coils are preformed in the shapes shown in Fig. 3.12, and are interconnected to form an armature winding. The coils span approximately a pole pitch, the distance between two consecutive poles.

The winding layout of a double-layer lap-wound armature is shown in Fig. 3.13. By tracing the circuit, as given below the winding lay out, it is seen that the entire winding is divided into four parallel paths. The layout of a double-

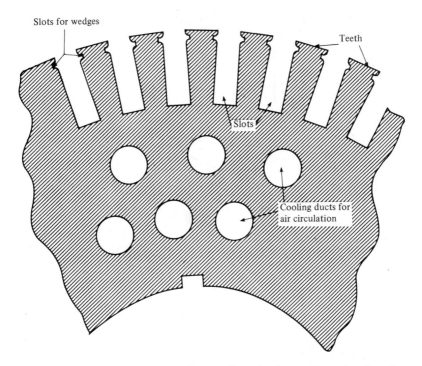

FIGURE 3-9. Portion of an armature lamination of a dc machine showing slots and teeth.

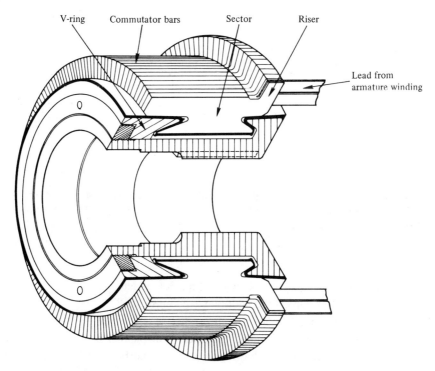

FIGURE 3-10. Details of commutator assembly.

3.4 CONSTRUCTIONAL DETAILS

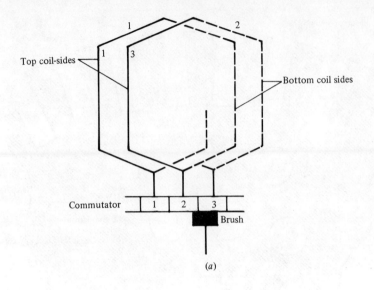

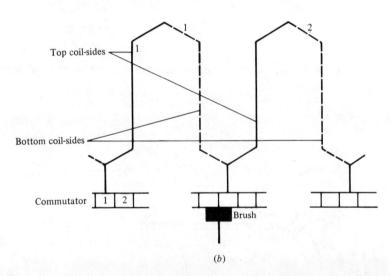

FIGURE 3-11. Elements of (a) lap winding, (b) wave winding. Odd-numbered conductors are at the top and even-numbered conductors are at the bottom of the slots.

layer wave winding is shown in Fig. 3.14. Here we have shown a simplex winding. For duplex and multiplex windings, see Reference 7. In a simplex lap winding the number of paths in parallel a, is equal to the number of poles p, whereas in a wave winding the number of parallel paths is always two. An assembled armature is shown in Fig. 3.15, and Fig. 3.16 shows an assembled field pole.

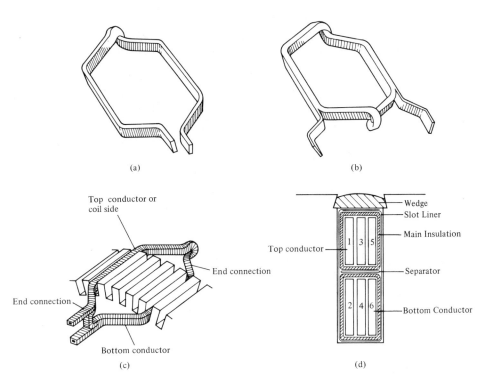

FIGURE 3-12. (a) Coil for a lap winding, made of a single bar; (b) multi-turn coil for wave winding; (c) a coil in a slot; (d) slot details, showing several coils arranged in two layers.

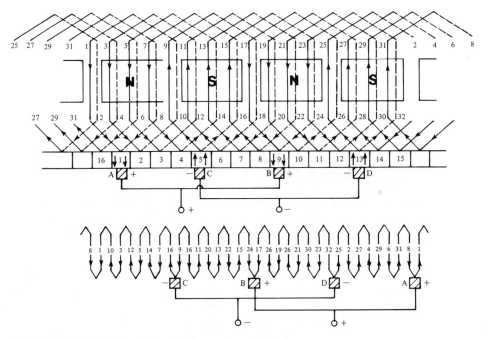

FIGURE 3-13. A 4-pole double-layer lap winding.

3.4 CONSTRUCTIONAL DETAILS

115

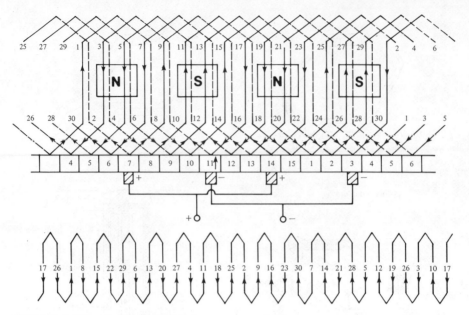

FIGURE 3-14. A 4-pole double-layer wave winding.

In addition to the armature and field windings, commutating poles and compensating windings are also found on large dc machines. These are shown in Fig. 3.17 and are used essentially to improve the performance of the machine, as we shall see later.

FIGURE 3-15. Armature of a 2000 kW 450 rpm dc generator (Courtesy of Brown Boveri Company).

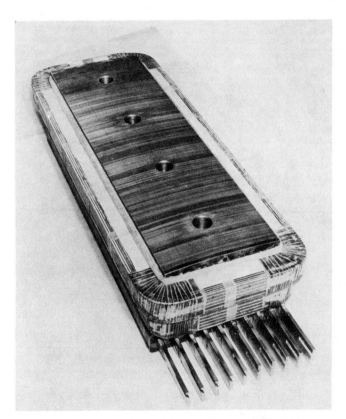

FIGURE 3-16. Field pole of a 2550 kW dc motor (Courtesy of Brown Boveri Company).

3.4 CONSTRUCTIONAL DETAILS

FIGURE 3-17. Stator of a 1030 kW dc motor showing interpoles and compensating bars (Courtesy of Brown Boveri Company).

3.5 CLASSIFICATION ACCORDING TO FORMS OF EXCITATION

Conventional dc machines that have a set of field windings and armature windings can be classified, on the basis of mutual electrical connections between the field and armature windings, as follows:

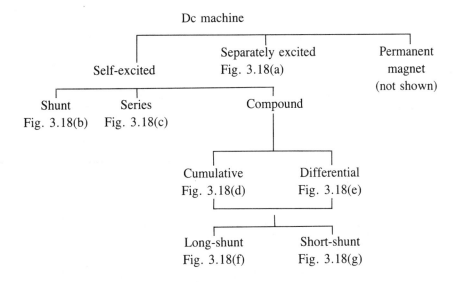

These interconnections of field and armature windings essentially determine the machine operating characteristics.

In a separately excited machine, shown in Fig. 3.18(a), there is no electrical interconnection between the field and the armature windings. On the other hand, the field winding is connected to the armature winding in a self-excited machine. A parallel connection between the field and armature windings results in the shunt machine of Fig. 3.18(b). The series machine has the field and the armature windings connected in series as in Fig. 3.18(c). A compound machine has both shunt and series field windings in addition to the armature winding. If the relative polarities of the shunt and series field windings are additive, as illustrated in Fig. 3.18(d), we obtain a cumulative compound machine. Notice from Fig. 3.18(d), that the two fields are shown to produce magnetic fluxes in the same direction. In a differential compound machine, the series field is in opposition to the shunt field, implying that the respective resulting fluxes are in opposition, as shown in Fig. 3.18(e). A differential or cumulative compound machine may have a long-shunt connection, in which case the shunt field is across the armature–series field combination, as given in Fig. 3.18(f). Figure 3.18(g) shows a short-shunt connection, where the shunt field is directly across the armature. Finally, in permanent magnet machine, we do not have the field winding and

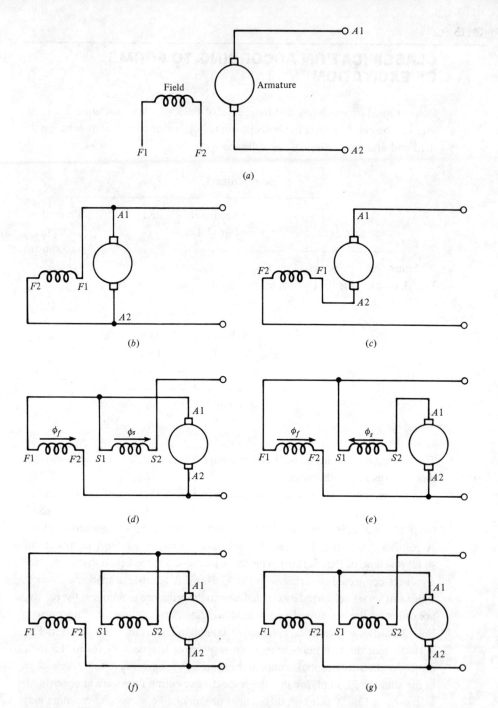

FIGURE 3-18. Classification of dc machines. (a) separately excited, (b) shunt, (c) series, (d) cumulative compound, (e) differential compound, (f) long-shunt, (g) short-shunt.

the necessary magnetic flux is provided by the permanent magnet. Such dc machines are generally of fractional-horsepower rating.

3.6 PERFORMANCE EQUATIONS

The three quantities of greatest interest in evaluating the performance of a dc machine are (1) the induced electromotive force (emf), (2) the electromagnetic torque developed by the machine, and (3) the speed corresponding to (1) and/or (2). We will now derive the equations that enable us to determine these quantities.

EMF Equation. The emf equation yields the emf induced in the armature of a dc machine. The derivation follows directly from Faraday's law of electromagnetic induction, according to which the emf induced in a moving conductor is the flux cut by the conductor per unit time.

Let us define the following symbols:

Z = number of active conductors on the armature

a = number of parallel paths in the armature winding

p = number of field poles

φ = flux per pole

n = speed of rotation of the armature, rpm

Then, with reference to Fig. 3.19,

flux cut by one conductor in one rotation = φp

flux cut by one conductor in n rotations = $\varphi n p$

flux cut per second by one conductor = $\dfrac{\varphi n p}{60}$

number of conductors in series = $\dfrac{Z}{a}$

flux cut per second by Z/a conductors = $\dfrac{\varphi n p}{60} \dfrac{Z}{a}$

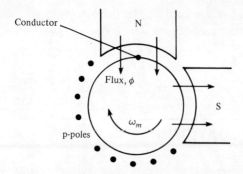

FIGURE 3-19. A conductor rotating at a speed ω_m in the field of p-poles.

Hence

emf induced in the armature winding

$$= E = \frac{\varphi nZ}{60} \times \frac{p}{a} \quad \text{V} \tag{3.7}$$

Equation (3.7) is known as the *emf equation* of a dc machine. In (3.7) we have separated p/a from the other terms because p and a are related to each other for the two types of windings.

EXAMPLE 3.2

Determine the voltage induced in the armature of a dc machine running at 1750 rpm and having four poles. The flux per pole is 25 mWb, and the armature is lap-wound with 728 conductors.

SOLUTION

Since the armature is lap-wound, $p = a$, and (3.7) becomes

$$E = \frac{\varphi nZ}{60} = \frac{25 \times 10^{-3} \times 1750 \times 728}{60} = 530.8 \text{ V}$$

Torque Equation. The mechanism of torque production in a dc machine has been considered earlier (see Fig. 3.6), and the electromagnetic torque developed by the armature can be evaluated either from the Bll rule (see Section 1.1) or from (3.1). Here we choose the latter approach. Regardless of the approach, it is clear that for torque production we must have a current through the armature, as this current interacts with the flux produced by the field winding. Let I_a be the armature current and E the voltage induced in the armature. Thus the power at the armature electrical port (see Fig. 3.1) is EI_a. Assuming that this entire electrical power is transformed into mechanical form, we rewrite (3.1) as

$$EI_a = T_e \omega_m \tag{3.8}$$

where T_e is the electromagnetic torque developed by the armature and ω_m is its angular velocity in rad/s. The speed n (in rpm) and ω_m (in rad/s) are related by

$$\omega_m = \frac{2\pi n}{60} \tag{3.9}$$

Hence from (3.7) to (3.9) we obtain

$$\frac{\omega_m}{2\pi} \varphi Z \frac{p}{a} I_a = T_e \omega_m$$

which simplifies to

$$T_e = \frac{Zp}{2\pi a} \varphi I_a \tag{3.10}$$

which is known as the *torque equation*. An application of (3.10) is illustrated by the next example.

EXAMPLE 3.3
A lap-wound armature has 576 conductors and carries an armature current of 123.5 A. If the flux per pole is 20 mWb, calculate the electromagnetic torque developed by the armature.

SOLUTION
For lap winding, we have $p = a$. Substituting this and other given numerical values in (3.10) yields

$$T_e = \frac{576}{2\pi} \times 0.02 \times 123.5 = 226.4 \text{ N-m}$$ ∎

EXAMPLE 3.4
If the armature of Example 3.3 rotates at an angular velocity of 150 rad/s, what is the induced emf in the armature?

3.6 PERFORMANCE EQUATIONS

SOLUTION

To solve this problem we use (3.8) rather than (3.7). From Example 3.3, we have $I_a = 123.5$ A and $T_e = 226.4$ N-m. Hence (3.8) gives

$$E \times 123.5 = 226.4 \times 150$$

or

$$E = 275 \text{ V} \qquad \blacksquare$$

Speed Equation and Back EMF. The emf and torque equations discussed above indicate that the armature of a dc machine, whether operating as a generator or as a motor, will have an emf induced while rotating in a magnetic field, and will develop a torque if the armature carries a current. In a generator, the induced emf is the internal voltage available from the generator. When the generator supplies a load, the armature carries a current and develops a torque. This torque opposes the prime mover (such as a diesel engine) torque.

In motor operation, the developed torque of the armature supplies the load connected to the shaft of the motor and the emf induced in the armature is termed the *back emf*. This emf opposes the terminal voltage of the motor.

Referring to Fig. 3.20, which shows the equivalent circuit of a separately excited dc motor running at speed n while taking an armature current I_a at a voltage V_t. From this circuit we have

$$V_t = E + I_a R_a \qquad (3.11)$$

Substituting (3.7) in (3.11), putting $k_1 = Zp/60a$, and solving for n, we get

$$n = \frac{V_t - I_a R_a}{k_1 \varphi} \qquad (3.12)$$

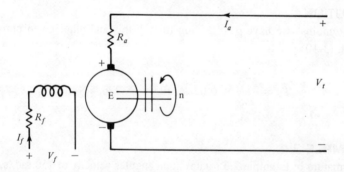

FIGURE 3-20. Equivalent circuit of a separately-excited motor.

This equation is known as the *speed equation*, as it contains all the factors that affect the speed of a motor. In a later section we shall consider the influence of these factors on the speed of dc motors. For the present, let us focus our attention on k_1 and φ. The term k_1 replacing $Zp/60a$ is a design constant in the sense that once the machine has been built, Z, p, and a cannot be altered. The magnetic flux φ is controlled primarily by the field current I_f (Fig. 3.20). If the magnetic circuit is unsaturated, φ is directly proportional to I_f. Thus we may write

$$\varphi = k_f I_f \tag{3.13}$$

where k_f is a constant. We may combine (3.12) and (3.13) to obtain

$$n = \frac{V_t - I_a R_a}{k I_f} \tag{3.14}$$

where $k = k_1 k_f$ = a constant. This form of the speed equation is more meaningful because all the quantities in (3.14) can be conveniently measured. [In contrast, in (3.12), it is very difficult to measure φ.]

EXAMPLE 3.5

A 250-V shunt motor has an armature resistance of 0.25 Ω and a field resistance of 125 Ω. At no-load the motor takes a line current of 5.0 A while running at 1200 rpm. If the line current at full-load is 52.0 A, what is the full-load speed?

SOLUTION

The motor equivalent circuit is shown in Fig. 3.21. The field current, $I_f = 250/125 = 2.0$ A.

At no load:

Armature current, $I_a = 5.0 - 2.0 = 3.0$ A

Back emf, $E_1 = V_t - I_a R_a = 250 - 3 \times 0.25 = 249.25$ V

Speed, $N_1 = 1200$ rpm (given)

At full-load:

Armature current $= 52.0 - 2.0 = 50.0$ A

Back emf, $E_2 = 250 - 50 \times 0.25 = 237.5$ V

Speed, $N_2 =$ unknown

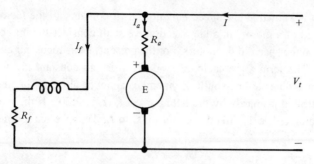

FIGURE 3-21. Equivalent circuit of a shunt motor.

Now,

$$\frac{N_2}{N_1} = \frac{E_2}{E_1} = \frac{237.5}{249.25}$$

Hence

$$N_2 = \frac{237.5}{249.25} \times 1200 = 1143 \text{ rpm}$$ ∎

3.7 ARMATURE REACTION

In the discussions so far we have assumed no interaction between the fluxes produced by the field windings and by the current-carrying armature windings. In reality, however, the situation is quite different. Consider the two-pole machine shown in Fig. 3.22(a). If the armature does not carry any current (that is, when the machine is on no-load), the air-gap field takes the form shown in Fig. 3.22(b). The geometric neutral plane and magnetic neutral plane (GNP and MNP) respectively, are coincident. (*Note:* Magnetic flux lines enter the MNP at right angles.) Noting the polarities of the induced voltages in the conductors, we see that the brushes are located at the MNP for maximum voltage at the brushes. We now assume that the machine is on "load" and that the armature carries current. The direction of flow of current in the armature conductors depends on the location of the brushes. For the situation shown in Fig. 3.22(b), the direction of the current flow is the same as the direction of the induced voltages. In any event, the current-carrying armature conductors produce their own magnetic fields, as shown in Fig. 3.22(c), and the air-gap field is now the resultant of the fields due to the field and armature windings. The resultant air-

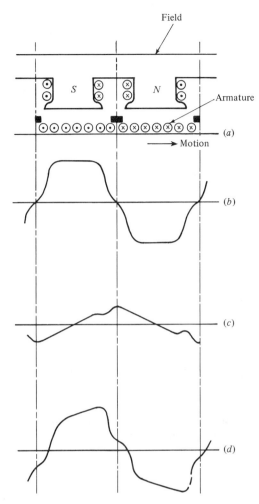

FIGURE 3-22. Air-gap fields in a dc machine: (a) a two-pole machine, showing armature and field MMF's; (b) flux-density distribution due to field MMF; (c) flux-density distribution due to armature MMF; (d) resultant flux-density distribution [curve (b) + curve (c)].

gap field is thus distorted and takes the form shown in Fig. 3.22(d). The interaction of the fields due to the armature and field windings is known as *armature reaction*. As a consequence of armature reaction the air-gap field is distorted and the MNP is no longer coincident with the GNP. For maximum voltage at the terminals, the brushes have to be located at the MNP. Thus one undesirable effect of armature reaction is that the brushes must be shifted constantly, since the shift of the MNP from the GNP depends on the load (which is presumably always changing). The effect of armature reaction can be analyzed in terms of cross-magnetization and demagnetization, as shown in Fig. 3.23(a). We just

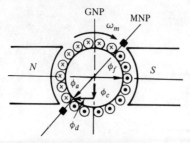

ϕ_a = Flux due to armature MMF
ϕ_c = Flux due to cross-magnetization
ϕ_d = Flux due to demagnetization
ϕ_f = Flux due to field MMF

(a)

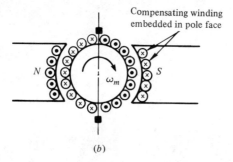

(b)

FIGURE 3-23. (a) Armature reaction resolved into cross and demagnetizing components; (b) neutralization of cross-magnetizing component by compensating winding.

mentioned the effect of cross-magnetization resulting in the distortion of the air-gap field and requiring the shifting of brushes according to the load on the machine. The effect of demagnetization is to weaken the air-gap field. All in all, therefore, armature reaction is not a desirable phenomenon in a machine.

The effect of cross-magnetization can be neutralized by means of compensating windings, as shown in Fig. 3.23. These are conductors embedded in pole faces, connected in series with the armature windings and carrying currents in an opposite direction to that flowing in the armature conductors under the pole face (Fig. 3.23). Once cross-magnetization has been neutralized, the MNP does not shift with load and remains coincident with the GNP at all loads. The effect of demagnetization can be compensated for by increasing the mmf on the main field poles. By neutralizing the net effect of armature reaction, we imply that there is no "coupling" between the armature and field windings.

3.8
REACTANCE VOLTAGE AND COMMUTATION

In discussing the action of the commutator in Chapter 2, we indicated that the direction of flow of current in a coil undergoing commutation reverses by the time the brushes move from one commutator segment to the other. This is represented schematically in Fig. 3.24. The flow of current in coil *a* for three different instants is shown. We have assumed that the current fed by a segment is proportional to the area of contact between the brush and the commutator segment. Thus, for satisfactory commutation, the direction of flow of current in coil *a* must completely reverse [Fig. 3.24(a) and (c)] by the time the brush moves from segment 2 to segment 3. The ideal situation is represented by the straight line in Fig. 3.25 and may be termed straight-line commutation. Because

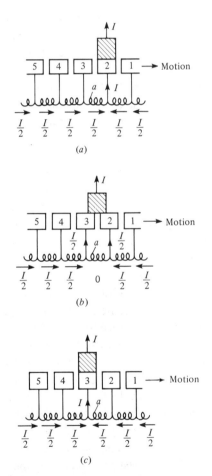

FIGURE 3-24. Coil *a* undergoing commutation.

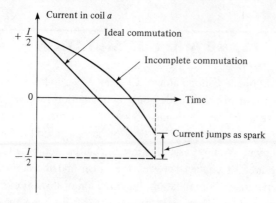

FIGURE 3-25. Commutation in coil a.

coil a has some inductance L, the change of current ΔI in a time Δt induces a voltage $L\,(\Delta I/\Delta t)$ in the coil. According to Lenz's law, the direction of this voltage, called *reactance voltage,* is opposite to the change (ΔI) that is causing it. As a result, the current in the coil does not completely reverse by the time the brush moves from one segment to the other. The balance of the "unreversed" current jumps over as a spark from the commutator to the brush, and thereby the commutator wears out because of pitting. This departure from ideal commutation is also shown in Fig. 3.25.

The directions of the (speed-) induced voltage, current flow, and reactance voltage are shown in Fig. 3.26(a). Note that the direction of the induced voltage depends on the direction of rotation of the armature conductors and on the direction of the air-gap flux. It is determined from the right-hand rule. Next, the direction of the current flow depends on the location of the brushes (or tapping points). Finally, the direction of the reactance voltage depends on the change in the direction of the current flow and is determined from Lenz's law. For the brush position shown in Fig. 3.26(a), observe that the reactance voltage retards the current reversal. If the brushes are advanced in the direction of rotation (for generator operation), we may notice, from Fig. 3.26(b), that the reactance voltage is in the same direction as the (speed-) induced voltage, and therefore the current reversal is not opposed. We may further observe that the coil undergoing commutation, being near the tip of the south pole, is under the influence of the field of a weak south pole. From this argument, we may conclude that commutation improves if we advance the brushes. But this is not a very practical solution. The same—perhaps better—results can be achieved if we keep the brushes at the GNP, or MNP, as in Fig. 3.26(a), but produce the "field of a weak south pole" by appropriately winding and connecting an auxiliary field winding, as shown in Fig. 3.26(c). The poles producing the desired field for better commutation are known as *commutating poles* or *interpoles*.

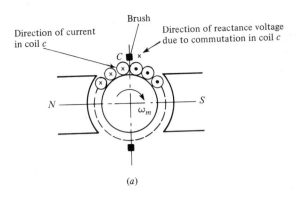

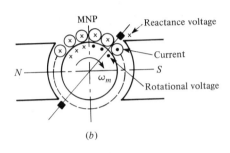

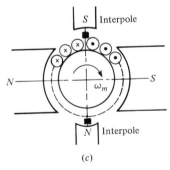

FIGURE 3-26. Reactance voltage and its neutralization: (a) reactance voltage and current in coil c, rotational voltage $\simeq 0$; (b) reactance voltage, rotational voltage, and current in coil c; (c) interpoles.

3.9
VOLTAGE BUILDUP IN A SHUNT GENERATOR

Saturation plays a very important role in governing the behavior of dc machines. It is extremely difficult to take into account the effects of saturation in the dynamical equations of motion. For the time being, let us consider qualitatively the consequences of saturation on the operation of a self-excited shunt generator.

A self-excited shunt machine is shown in Fig. 3.21. We write the steady-state equations for the operation of the machine as a generator. From the circuit shown in Fig. 3.21 we have

$$V_t = R_f I_f$$

and

$$E = V_t + I_a R_a = I_f R_f + I_a R_a$$

These equations are represented by the straight lines shown in Fig. 3.27(a). Notice that the voltages V_t and E will keep building up and no equilibrium point

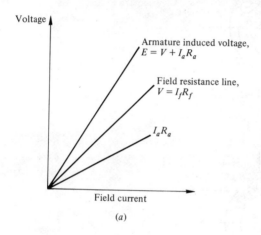

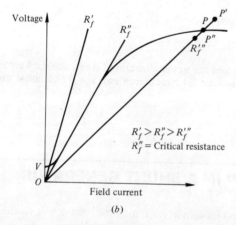

FIGURE 3-27. No-load characteristic of shunt generator: (a) no stable operating point for the shunt generator; (b) stable no-load voltage of a shunt generator.

can be reached. On the other hand, if we include the effect of saturation, as in Fig. 3.27(b), point P defines an equilibrium, because at this point the field-resistance line intersects the saturation curve. A deviation from P to P' or P'' would immediately show that at P' the voltage drop across the field is greater than the induced voltage, which is not possible; and at P'' the induced voltage is greater than the field-circuit voltage drop, which is not possible either.

The small voltage OV shown in Fig. 3.27(b) results from the residual magnetism of the field poles. Evidently, without this remanent flux the shunt generator will not build up any voltage. Also shown in Fig. 3.27(b) is the critical resistance. A field resistance greater than the critical resistance (for a given speed) would not let the shunt generator build up any appreciable voltage. Finally, we should ascertain that the polarity of the field winding is such that a current through it produces a flux that aids the residual flux. If it does not, the two fluxes tend to neutralize each other and the machine voltage will not build up. To summarize, the conditions for voltage buildup in a shunt generator are the presence of residual flux, field-circuit resistance less than the critical resistance, and appropriate polarity of the field winding.

3.10 GENERATOR CHARACTERISTICS

No-load and load characteristics of dc generators are usually of interest in determining their potential applications. Of the two, load characteristics are of greater importance. As the names imply, no-load and load characteristics correspond, respectively, to the behavior of the machine when it is supplying no power (open-circuited, in the case of a generator) and when it is supplying power to an external circuit.

The only no-load (or open-circuit) characteristics that are meaningful are those of the shunt and separately excited generators. We have discussed the no-load characteristic of a shunt generator as a voltage buildup process in the preceding section. For the separately excited generator, the no-load characteristic corresponds to the magnetization, or saturation, characteristic—variation of E (or V_t) under open-circuit condition as a function of the field current, I_f. This characteristic is illustrated in Fig. 3.28, where the symbols are as shown in Fig. 3.20.

Turning now to the load characteristics, we define it as a variation of the terminal voltage as a function of load current supplied by the generator. This characteristic is fairly straightforward to obtain if we can identify the causes of voltage drops in dc generators. The main causes of the voltage drop in generators are:

1. *Armature resistance drop:* This is an $I_a R_a$ drop due to the resistance of the armature.

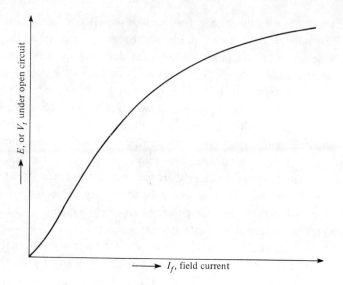

FIGURE 3-28. No-load characteristic of a separately-excited generator.

2. *Brush contact drop:* The mechanical contact between the brushes and the commutator offers an electrical resistance. Consequently, when a current flows through the brush, a voltage drop occurs. Usually, this voltage drop is taken as a constant (of 2 V).
3. *Armature reaction voltage drop:* From Section 3.7 we recall that armature reaction has a demagnetizing component, which opposes the main field mmf, resulting in a reduction of flux. The reduced flux will, in turn, reduce the armature induced emf and hence the terminal voltage.
4. Cumulative effects of (1) to (3) in self-excited (shunt, series, and compound) generators further lower the terminal voltage.

In view of the above, let us refer to Fig. 3.29, which shows the load characteristic of a separately excited dc generator. Notice that the terminal voltage on load differs from the no-load voltage by the three voltage drops mentioned in (1) to (3). The load characteristics of self-excited generators are shown in Fig. 3.30. The shunt generator has a characteristic similar to that of a separately excited generator, except for the cumulative effect mentioned in (4). If the shunt generator is loaded beyond a certain point, it breaks down in that the terminal voltage collapses. In a series generator, the load current flows through the field winding. This implies that the field flux, and hence the induced emf, increases with the load until the core begins to saturate magnetically. A load beyond a certain point would also result in a collapse of the terminal voltage of the series generator. Compound generators have combined characteristics of shunt and series generators. In a differential compound generator, shunt and series fields are in opposition. Hence the terminal voltage drops very rapidly with the load.

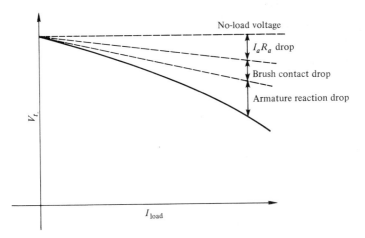

FIGURE 3-29. Load characteristic of a separately-excited generator.

On the other hand, cumulative compound generators have shunt and series fields aiding each other. The two field mmf's could be adjusted such that the terminal voltage on full-load is less than the no-load voltage, as in an under-compound generator; or the full-load voltage may be equal to the no-load voltage, as in a flat-compound generator. Finally, the terminal voltage on full-load may be greater than the no-load voltage, as in an over-compound generator.

EXAMPLE 3.6

A 50-kW 250-V short-shunt compound generator has the following data: $R_a = 0.06\ \Omega$ and $R_f = 125\ \Omega$. Calculate the induced armature emf at rated load and terminal voltage. Take 2 V as the total brush-contact drop.

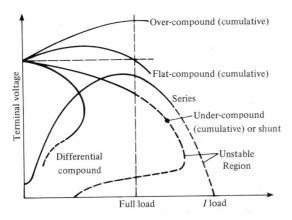

FIGURE 3-30. Load characteristics of dc generators.

3.10 GENERATOR CHARACTERISTICS

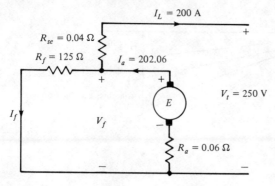

FIGURE 3-31. Example 3.6.

SOLUTION

The equivalent circuit of the generator is shown in Fig. 3.31, from which

$$I_t = \frac{50 \times 10^3}{250} = 200 \text{ A}$$

$$I_t R_{se} = (200)(0.04) = 8 \text{ V}$$

$$V_f = 250 + 8 = 258 \text{ V}$$

$$I_f = \frac{258}{125} = 2.06 \text{ A}$$

$$I_a = 200 + 2.06 = 202.06 \text{ A}$$

$$I_a R_a = (202.06)(0.06) = 12.12 \text{ V}$$

$$E = 250 + 12.12 + 8 + 2 = 272.12 \text{ V} \quad \blacksquare$$

3.11

MOTOR CHARACTERISTICS

Among the various characteristics of dc motors, their torque–speed characteristics are most important from a practical standpoint. The torque and speed equations derived earlier govern the motor characteristics. From these equations (and after accounting for magnetic saturation) it follows that the shunt, series, and cumulative compound motors have the torque–speed characteristics of the forms shown in Fig. 3.32. The governing equations also yield the motor speed–current characteristics of Fig. 3.33.

In Fig. 3.32 we have also shown the developed power versus speed for shunt and series motors. As we recall from Section 3.1, developed power is simply

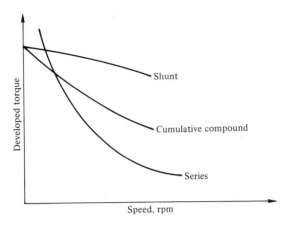

FIGURE 3-32. Torque–speed characteristics of dc motors.

the product of developed torque (in N-m) and the angular speed (in rad/s). From Fig. 3.32 and 3.33 it is clear that a series motor on no-load will run at a dangerously high speed. Similarly, it follows from the speed equation that loss of field excitation in a shunt motor will result in an overspeeding of the motor.

3.12
STARTING AND CONTROL OF MOTORS

In addition to certain operational conveniences, the basic requirements for satisfactory starting of a dc motor are (1) sufficient starting torque, and (2) armature

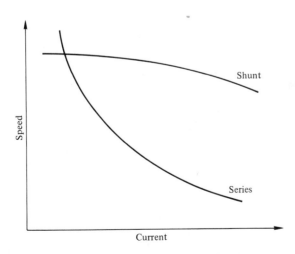

FIGURE 3-33. Speed–current characteristics.

current, within safe limits, for successful commutation and for preventing the armature from overheating. The second requirement is obvious from the speed equation, according to which the armature current, I_a, is given by $I_a = V_t/R_a$ when the motor is at rest (n or $\omega_m = 0$). A typical 50-hp 230-V motor having an armature resistance of 0.05 Ω, if connected across 230 V, shall draw 4600 A of current. This current is evidently too large for the motor, which might be rated to take 180 A on full-load. Commonly, double the full-load current is allowed to flow through the armature at the time of starting. For the motor under consideration, therefore, an external resistance

$$R_{ext} = \left(\frac{230}{2 \times 180}\right) - 0.05 = 0.59 \text{ Ω}$$

must be inserted in series with the armature to limit I_a within double the rated value.

In practice, the necessary starting resistance is provided by means of a starter. Typical three-point and four-point starters are shown in Fig. 3.34. Notice that at the time of starting, the resistance R_{ext} comes in series with the armature. As

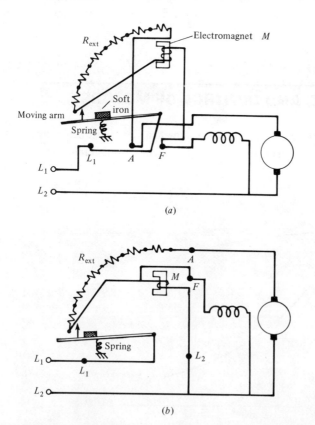

FIGURE 3-34. (a) a three-point starter; (b) a four-point starter.

the motor speeds up this resistance is cut out in steps. When the entire resistance is cut out, the starter arm is held by the electromagnet M. When the supply is turned off the starter arm is pulled back to the "off" position by the spring. Notice from Fig. 3.34(a) that if the field is opened, the electromagnet M is deenergized and the arm is pulled back to the "off" position. This provides field-failure protection. But if a large resistance is connected in series with the field circuit, the electromagnet may fail to hold the arm. On the other hand, there is no field-failure protection in a four-point starter, as is clear from Fig. 3.34(b). The electromagnet current is independent of the field current.

The three- and four-point starters just mentioned are old-fashioned starters. Modern starters are of the automatic pushbutton type. The switching of resistances is made automatically by contactors. Numerous versions of automatic starters for dc motors are commercially available. To illustrate the principle of operation of pushbutton starters, we consider only one type here, as shown in Fig. 3.35. Here, the contactors and the corresponding relays have the same numbers. For instance, a current in relay coil 1 will operate contactor 1. If this contactor is normally open, it will be closed when a predetermined current flows through the coil, and vice versa. Let us consider the sequence of operation. The start pushbutton is normally open and the stop pushbutton is normally closed. To start the motor, we push the start button, which makes a current to flow relay coil 1. This closes contactor 1 and current begins to flow through relay 2 (which is a current relay for over-current protection) and the armature. The

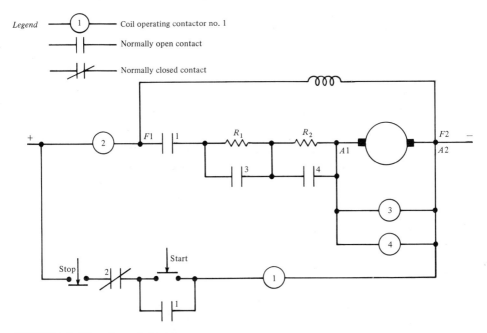

FIGURE 3-35. A push-button starter.

armature current is limited by resistances R_1 and R_2 and the armature resistance. As the motor picks up speed, it develops a back emf and thereby a voltage is applied across relays 3 and 4. Relay 4 operates at a higher voltage than the operating voltage of relay 3. At a certain value of the back emf, relay 3 operates, contactor 3 closes, and R_1 is short-circuited. At a higher speed, relay 4 operates to close contactor 4 and short-circuit the resistance R_2. To stop the motor, we push the stop button to deenergize relay 1, thereby opening contactor 1 and open-circuiting the armature.

From the speed equation of a dc motor, it follows that the speed of the motor can be varied by varying (1) the field-circuit resistance to control I_f and hence the field flux, (2) the armature circuit resistance, and (3) the terminal voltage. Let us now consider the scope of each of these three methods. In method (1), an external variable resistance is connected in the field circuit. When this resistance is set at zero, the field current is limited only by the field resistance. Corresponding to this field current, we obtain a motor speed, n_1. Now, as the externally inserted resistance increases, the field current decreases and the motor speed increases (in accordance with the speed equation) to a corresponding speed n_2, where $n_2 > n_1$. In other words, by using an external resistance in series with the motor field, we can only increase the motor speed (from a minimum speed n_1). By inserting a resistance in series with the armature, as in method (2), we can reduce the voltage across the armature and hence decrease the motor speed. Because the armature current is of a relatively large magnitude, compared to the field current, method (2) is a wasteful method (as shown by Example 3.8). Method (3) can be efficiently used either to increase or decrease the speed of the motor, but is feasible only if a variable voltage source is available. Method (3) in conjunction with method (1) constitutes the *Ward–Leonard system*, which is capable of providing a wide variation of speed in both forward and reverse directions. This method is presented in a simplified form in Example 3.9.

EXAMPLE 3.7

A 230-V shunt motor has an armature resistance of 0.05 Ω and a field resistance of 75 Ω. The motor draws 7 A of line current while running light at 1120 rpm. The line current at a certain load is 46 A. (a) What is the motor speed at this load? (b) At this load, if the field-circuit resistance is increased to 100 Ω, what is the new speed of the motor? Assume the line current to remain unchanged.

SOLUTION
(a) On no-load,

$$N_0 = 1120 \text{ rpm (given)}$$

$$I_f = \frac{230}{75} = 3.7 \text{ A}$$

$$I_a = 7 - 3.07 = 3.93 \text{ A}$$

The speed equation gives

$$1120 = \frac{230 - 3.93 \times 0.05}{3.07k}$$

or

$$k = 0.0668$$

On load (with $R_f = 75\ \Omega$):

$$I_f = 3.07\text{ A}$$

$$I_a = 46 - 3.07 = 42.93\text{ A}$$

$$n = \frac{230 - 42.93 \times 0.05}{3.07 \times 0.0668} = 1111\text{ rpm}$$

(b) On load (with $R_f = 100\ \Omega$):

$$I_f = \frac{230}{100} = 2.3\text{ A}$$

$$I_a = 46 - 2.3 = 43.7\text{ A}$$

$$n = \frac{230 - 43.7 \times 0.05}{2.3 \times 0.0668} = 1483\text{ rpm}$$

■

EXAMPLE 3.8
Refer to part (a) of Example 3.7. The no-load conditions remain unchanged. On load, the line current remains at 46 A, but a 0.1-Ω resistance is inserted in the armature. Determine the speed of the motor and the power dissipated in the 0.1-Ω resistance.

SOLUTION
In this case we have (from Example 3.7)

$$I_f = 3.07\text{ A}$$

$$I_a = 42.93\text{ A}$$

$$k = 0.0668$$

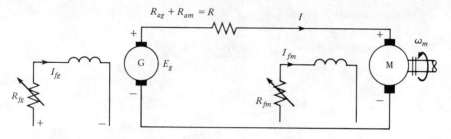

FIGURE 3-36. A Ward-Leonard System.

Thus, with $R_a = 0.05 + 0.1 = 0.15 \, \Omega$,

$$I_a^2(0.1) = 42.93^2 \times 0.1 = 184.3 \text{ W}$$

$$n = \frac{230 - 42.93 \times 0.15}{3.07 \times 0.0688} = 1058 \text{ rpm}$$

EXAMPLE 3.9

The system shown in Fig. 3.36 is called the *Ward–Leonard system* for controlling the speed of a dc motor. Discuss the effects of varying R_{fg} and R_{fm} on the motor speed.

SOLUTION

Increasing R_{fg} decreases I_{fg} and hence E_g. Thus the motor speed will decrease. The opposite will be true if R_{fg} is decreased.

Increasing R_{fm} will increase the speed of the motor, as shown in Example 3.7. Decreasing R_{fm} will result in a decrease of the speed.

3.13

LOSSES AND EFFICIENCY

Besides the voltage–amperage and speed–torque characteristics, the performance of a dc machine is measured by its efficiency:

$$\text{efficiency} = \frac{\text{power output}}{\text{power input}} = \frac{\text{power output}}{\text{power output} + \text{losses}}$$

$$= \frac{\text{power input} - \text{losses}}{\text{power input}} \quad (3.15)$$

Efficiency may, therefore, be determined either from load tests or by determination of losses. The various losses are classified as follows:

1. *Electrical:* (a) Copper losses in various windings, such as the armature winding and different field windings; and (b) loss due to the contact resistance of the brush (with the commutator).
2. *Magnetic:* These are the iron losses and include the hysteresis and eddy-current losses in the various magnetic circuits, primarily the armature core and pole faces.
3. *Mechanical:* These include the bearing-friction, windage, and brush-friction losses.
4. *Stray-load:* These are other load losses not covered above. They are taken as 1 percent of the output (as a rule of thumb). Another recommendation often followed is to take the stray-load loss as 28 percent of the core loss at the rated output.

The power flow in a dc generator or motor is represented in Fig. 3.37, in which the symbols are as follows: V_t = terminal voltage, V; I_a = armature current, A; E = back or induced emf in the armature, V; V_f = voltage across the field winding, V; I_f = current through the field winding, A; T_e = electromagnetic torque developed by the armature, N-m; T_s = torque available at the shaft, N-m; P_{elec} = electrical power, W; P_{SL} = stray-load loss, W; P_{mech} = rotational mechanical loss, W; P_{mag} = magnetic losses, W; and ω_m = mechanical angular velocity, rad/s.

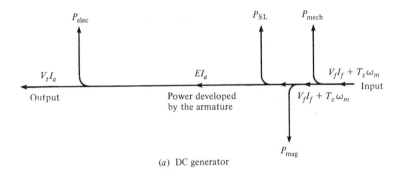

(a) DC generator

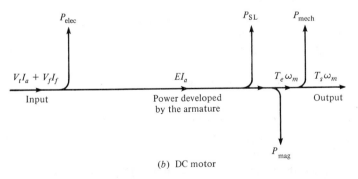

(b) DC motor

FIGURE 3-37. Power flow in separately-excited (a) dc generator (b) dc motor.

3.13 LOSSES AND EFFICIENCY

EXAMPLE 3.10

A 10-hp 230-V shunt motor takes a full-load line current of 40 A. The armature and field resistances are 0.25 Ω and 230 Ω, respectively. The total brush-contact drop is 2 V and the core and friction losses are 380 W. Calculate the efficiency of the motor. Assume that stray-load loss is 1 percent of rated output.

SOLUTION

input = (40)(230) = 9200 W

field-resistance loss = $\dfrac{230^2}{230}$ (230) = 230 W

armature-resistance loss = $(40 - 1)^2 (0.25)$ = 380 W

core loss and friction loss = 380 W

brush-contact loss = (2)(39) = 78 W

stray-load loss = $\dfrac{10}{100} \times 746$ = 75 W

total losses = 1143 W

power output = 9200 − 1143 = 8057 W

$$\text{efficiency} = \dfrac{8057}{9200} = 87.6\%$$

■

3.14

TESTS ON DC MACHINES

Tests are performed on dc machines for the following purposes:

1. To obtain no-load and magnetization characteristics.
2. To obtain load characteristics, including the determination of losses and efficiency.
3. To evaluate temperature rise.
4. To assess commutation.

No-Load Tests. To determine the magnetization characteristic, the machine under test is driven as a generator at a constant speed. The field is excited separately and the armature is open-circuited. A plot of the open-circuit armature voltage as a function of the field current gives the magnetization characteristic. A typical magnetization characteristic is shown in Fig. 3.28.

On no-load, the input power is the sum of magnetic, mechanical, and electrical losses in the field and the armature. If we run the machine as a motor on

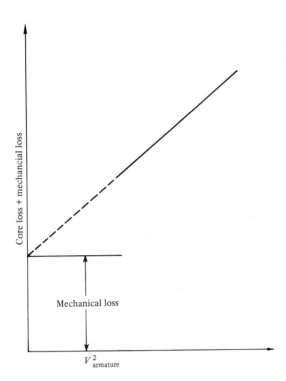

FIGURE 3-38. Separation of mechanical loss.

no-load (at a constant speed by adjusting the armature voltage) and vary the field excitation, the input power will be the sum of a constant mechanical loss, a varying core loss, and electrical losses in the field and the armature. Subtracting the electrical losses from the input power gives the constant mechanical loss and the variable core loss. The core loss is approximately proportional to the square of the armature voltage. If we plot the sum of the core loss and the mechanical loss versus (armature voltage)^2, we will obtain a straight line. The intercept of this straight line (by extrapolation) with the vertical (Fig. 3.38) gives the mechanical losses.

Load Tests. Small dc machines may be tested on load by loading them directly by a mechanical brake or a dynamometer. Larger machines are evaluated for their characteristics by segregation of losses (see Example 3.10). Direct testing of large machines is uneconomical and often not feasible. In such cases, back-to-back tests may be performed if two identical machines are available. This test is rather specialized and we will not consider it here.

Temperature Rise. Losses in electric machines are dissipated as heat and result in a temperature rise in the machine. The temperature rise depends on the rate of heat generation, heat transfer, and thermal capacity. The rating of a machine is determined by the temperature rise. Heat-run tests on small machines can be performed by loading them as generators. Temperatures within the machine can be monitored by strategically located thermocouples. Large machines are tested for temperature rise by the back-to-back test.

Commutation. There are a number of tests conducted to assess the performance of the commutator of a dc machine. These include a check of the brush location, contact drop, armature faults, and black-band test. A simple test to check the brush location is to find by means of probes the two adjacent commutator segments between which no instantaneous voltage is observable when the field is suddenly turned on and turned off.

The contact drop is illustrated by Fig. 3.39, and checks the uniformity of current density under the brush. If the voltage drop between the commutator and the brush at a few points along its width gives a straight line, we have a correct condition for commutation. Under- and over commutation are also shown in Fig. 3.39.

Faults in the armature winding can be detected by passing a current through the stationary armature and measuring the voltages between the segments around the commutator. If these voltages are equal, there is no fault. A low voltage indictes a short circuit and an open-circuited coil is indicated by zero voltage between all pairs of segments in the faulty path except the pair at the terminals of the faulty coil.

The black-band test is used to check the effectiveness of the commutating poles. In this test, for a range of fixed armature currents, the current in the commutating pole is reduced until sparking begins. The commutating pole current is then gradually increased until sparking occurs again. The results of this test are shown in Fig. 3.40. The range of sparkless commutation is called the *black band*. If the black band is asymmetrical at low armature currents, the brushes are incorrectly located; if tilted with increase of armature current, the machine is over- or under-commutated; and if very narrow at large armature

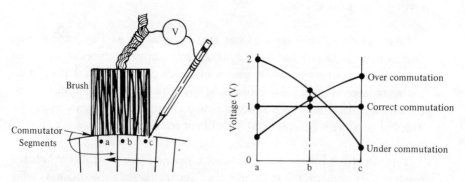

FIGURE 3-39. Contact-drop test.

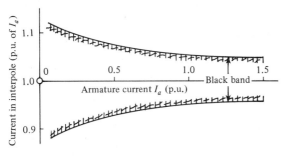

FIGURE 3-40. Black band.

currents, the commutating poles are saturated. The per unit current in Fig. 3.40 is the ratio of the actual current to the full-load armature current.

3.15
CERTAIN APPLICATIONS

Dc machines find applications in the following: electric traction and diesel-electric locomotives; large rolling mills; electrochemical plants and metal refining plants; elevators; earth-moving equipment; battery charging; ship, train, and aircraft auxiliaries; isolated experimental stations; excitors for synchronous machines (see Chapter 4); automatic control systems; wind-generating systems; and so on.

The choice of a dc motor for a specific application depends on the nature of the load. For instance, permanent-magnet dc motors are preferred for actuators requiring high pcak and steady power and fast response. They are basic drives in aircraft control systems. For traction loads, dc series motors are ideally suited. Series motors are used to drive cranes, hoists, and high-inertia loads. In contrast, the shunt motor is essentially a constant-speed motor. Its speed can be easily controlled by adjusting the field current or armature voltage. Hence it has numerous industrial applications, such as in driving pumps, compressors, punch presses, and so on.

Dc generators are used in the chemical industry for applications requiring electrolysis processing. Differential compound generators are used in welding.

3.16
PARALLEL OPERATION OF DC GENERATORS

Generators are operated in parallel for load sharing; that is when the demand exceeds the capacity of one generator, a second one is operated in parallel to provide an adequate supply for the load.

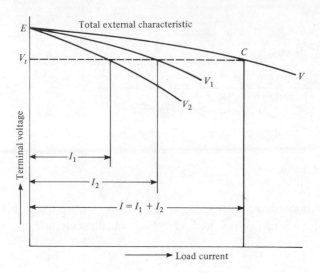

FIGURE 3-41. Load sharing between two shunt generators having the same no-load voltage.

Shunt Generators in Parallel. The load (or external) characteristics of two shunt generators are shown in Fig. 3.41. The field excitation of the two generators is so adjusted that their open-circuit terminal voltages are equal. For a terminal voltage V_t, the load current I is divided into I_1, delivered by the first generator, and I_2, supplied by the second generator. Clearly, $I = I_1 + I_2$. The generator with the more drooping characteristic takes the smaller share of the load. Because of their drooping characteristics, shunt generators inherently work well in parallel. There is no tendency for an unstable operation.

If the open-circuit voltages of the two generators are not equal, as shown in Fig. 3.42, then at very light loads the machine with the smaller open-circuit

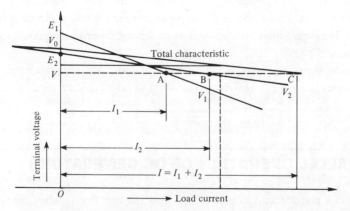

FIGURE 3-42. Load sharing between two shunt generators having unequal induced voltages.

voltage receives current (instead of supplying) and tends to act as a motor. Beyond point A in Fig. 3.42, both machines work as generators.

Load sharing between shunt generators can be obtained graphically, as illustrated in Fig. 3.41 and 3.42. On the other hand, if it is assumed that the terminal voltage of a generator varies linearly with the current supplied by the generator, the following relations hold:

$$V_0 = \frac{E_1 R_{a2} + E_2 R_{a1}}{R_{a1} + R_{a2}}$$

$$I_0 = \frac{E_1 - E_2}{R_{a1} + R_{a2}}$$

$$I_1 = \frac{E_1 R_{a2} + (E_1 - E_2)R}{R_{a1} R_{a2} + (R_{a1} + R_{a2})R}$$

$$I_2 = \frac{E_2 R_{a1} + (E_2 - E_1)R}{R_{a1} R_{a2} + (R_{a1} + R_{a2})R}$$

$$I = \frac{E_1 R_{a2} + E_2 R_{a1}}{R_{a1} R_{a2} + (R_{a1} + R_{a2})R}$$

where V_0 is the no-load terminal voltage of the two generators in parallel; I_0 the no-load circulating current; E_1 and E_2 the no-load induced voltages of generators 1 and 2 respectively; I_1 and I_2 the currents supplied by the two generators; R_{a1} and R_{a2} the respective armature circuit resistances; I the load current; and R the load resistance. The terminal voltage, V, on load is then given by

$$V = \frac{E_1 R_{a2} + E_2 R_{a1}}{(R_{a1} R_{a2}/R) + (R_{a1} + R_{a2})}$$

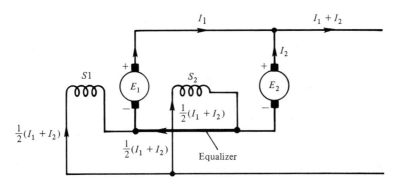

FIGURE 3-43. Series generators in parallel showing equalizer connections.

3.16 PARALLEL OPERATION

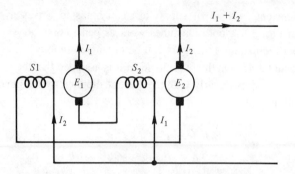

FIGURE 3-44. Series generators in parallel showing cross-connected series fields.

Series and Cumulative Compound Generators in Parallel.
Because series and cumulative compound generators do not have drooping characteristics, their parallel operation is inherently unstable. For their satisfactory operation in parallel either an equalizer connection (Fig. 3.43) must be used or the field windings must be cross-connected (Fig. 3.44). By the equalizer connection or the cross-connection, the two field windings are connected in parallel, and the flux in either machine is not directly proportional to its own armature current. If an equalizer connection is used, its resistance must be small compared to the resistances of the series-field windings.

QUESTIONS

3.1 Faraday's law of electromagnetic induction states that an emf is induced in a closed circuit if the magnetic flux linking the circuit changes with time. Can this law be adapted to obtain the emf induced in a conductor "cutting" magnetic flux? Explain.

3.2 Refer to Fig. 3.2. Let a current enter at the positive terminal. By applying the left-hand rule, determine the possible direction of rotation of the disk. Verify that the machine is capable of operating as a dc motor.

3.3 What is the purpose of a commutator in a conventional dc machine?

3.4 Why does a homopolar dc machine not need a commutator?

3.5 What is the difference between lap and wave windings? Which of the two windings is suitable for (a) high voltage? (b) high current?

3.6 Why is it necessary to make the armature of a dc machine of iron?

3.7 In a dc machine the armature core is laminated but not the pole core. Why?

3.8 Why is it necessary to laminate the pole faces, but not the pole cores, of a dc machine?

3.9 What are the basic functions of the following components of a dc machine: (a) armature winding; (b) field winding; (c) yoke; (d) field poles; (e) armature core; (f) slots and teeth; (g) interpoles; (h) compensating windings; and (i) brushes?

3.10 What is the relationship between the number of poles and the number of parallel paths in (a) a lap winding and (b) a wave winding of a dc armature?

3.11 How are dc machines classified according to the form of excitation?

3.12 What is meant by "back emf" in a dc motor?

3.13 What are various possible methods of speed control of a dc motor? What are the relative advantages and disadvantages of these methods?

3.14 What are the consequences of armature reaction in a dc generator?

3.15 How is armature reaction neutralized?

3.16 What is reactance voltage? How is it related to the process of commutation?

3.17 Explain the process of voltage buildup in a self-excited shunt generator. Would a shunt generator build up to a steady value of a voltage if the magnetic circuit of the machine did not saturate?

3.18 What is the significance of critical resistance in a self-excited shunt generator?

3.19 Why is it necessary to connect a resistance in series with the armature of a dc motor at the time of its starting?

3.20 What is the difference between the geometric neutral plane (GNP) and the magnetic neutral plane (MNP) of a dc machine?

PROBLEMS

3.1 The armature of a four-pole dc machine has 32 conductors. Draw the winding layout for a double-layer lap winding, and verify that the winding has four parallel paths.

3.2 The armature of a four-pole dc machine has 30 conductors. Draw the winding layout for a two-layer wave winding. Verify that the winding has two parallel paths.

3.3 The flux per pole of a generator is 75 mWb. The generator runs at 900 rpm. Determine the induced voltage if (a) the armature has 32 conductors connected as lap winding, or (b) the armature has 30 conductors connected as wave winding.

3.4 Determine the flux per pole of a six-pole generator required to generate 240 V at 500 rpm. The armature has 120 slots, with eight conductors per slot, and is lap connected.

3.5 If the armature current in the generator of Problem 3.4 is 25 A, what electromagnetic torque is developed?

3.6 The armature and field resistances of a 240-V shunt generator are 0.2 Ω and 200 Ω, respectively. The generator supplies a 9600-W load. Determine the induced voltage, taking 2 V as the total brush-contact voltage drop.

3.7 The armature, shunt field, and the series field resistances of a compound generator are 0.02 Ω, 80 Ω, and 0.03 Ω, respectively. The generator induced voltage is 510 V and the terminal voltage is 500 V. Calculate the power supplied to the load (at 500 V) if the generator has (a) a long-shunt connection and (b) a short-shunt connection.

3.8 A four-pole shunt-connected generator has a lap-connected armature with 728 conductors. The flux per pole is 25 mWb. If the generator supplies two hundred 110-V 75-W bulbs, determine the speed of the generator. The field and armature resistances are 110 Ω and 0.075 Ω, respectively.

3.9 A shunt generator delivers 50 A of current to a load at 110 V, at an efficiency of 85 percent. The total constant losses are 480 W, and the shunt-field resistance is 65 Ω. Calculate the armature resistance.

3.10 For the generator of Problem 3.9, plot a curve for efficiency versus armature current. At what armature current is the efficiency maximum?

3.11 A separately excited six-pole generator has a 30-mWb flux per pole. The armature is lap wound and has 534 conductors. This supplies a certain load at 250 V while running at 1000 rpm. At this load, the armature copper loss is 640 W. Calculate the load supplied by the generator. Take 2 V as the total brush-contact drop.

3.12 In a dc machine, the hysteresis and eddy-current losses at 1000 rpm are 10,000 W at a field current of 7.8 A. At 750-rpm speed and 7.8-A field current, the total iron losses become 6000 W. Assuming that the hysteresis loss is directly proportional to the speed and that the eddy-current loss is proportional to the square of the speed, determine the hysteresis and eddy-current losses at 500 rpm.

3.13 The saturation characteristic of a dc shunt generator is as follows:

Field Current (A)	1	2	3	4	5	6	7
Open-Circuit Voltage (V)	53	106	150	192	227	252	270

The generator speed is 900 rpm. At this speed, what is the maximum field-circuit resistance such that the self-excited shunt generator would not fail to build up?

3.14 To what value will the no-load voltage of the generator of Problem 3.13 build up, at 900 rpm, for a field-circuit resistance of 42 Ω?

3.15 A 240-V separately excited dc machine has an armature resistance of 0.25 Ω. The armature current is 56 A. Calculate the induced voltage for (a) generator operation and (b) motor operation.

3.16 The field and armature winding resistances of a 400-V dc shunt machine are 120 Ω and 0.12 Ω, respectively. Calculate the power developed by the armature if (a) the machine takes 50 kW while running as a motor, and (b) the machine delivers 50 kW while running as a generator.

3.17 The field and armature resistances of a 220-V series motor are 0.2 Ω and 0.1 Ω, respectively. The motor takes 30 A of current while running at 700 rpm. If the total iron and friction losses are 350 W, determine the motor efficiency.

3.18 A 400-V shunt motor delivers 15 kW of power at the shaft at 1200 rpm while drawing a line current of 62 A. The field and armature resistances are 200 Ω and 0.05 Ω, respectively. Assuming 1 V of contact drop per brush, calculate (a) the torque developed by the motor and (b) the motor efficiency.

3.19 A 400-V series motor, having an armature circuit resistance of 0.5 Ω, takes 44 A of current while running at 650 rpm. What is the motor speed for a line current of 36 A?

3.20 A 220-V shunt motor having an armature resistance of 0.2 Ω and a field resistance of 110 Ω takes 4 A of line current while running on no-load. When loaded, the motor runs at 1000 rpm while taking 42 A of current. Calculate the no-load speed.

3.21 The machine of Problem 3.20 is driven as a shunt generator to deliver a 44-kW load at 220 V. If the machine takes 44 kW while running as a motor, what is its speed?

3.22 A 220-V shunt motor having an armature resistance of 0.2 Ω and a field resistance of 110 Ω takes 4 A of line current while running at 1200 rpm on no-load. On load, the input to the motor is 15 kW. Calculate (a) the speed, (b) the developed torque, and (c) the efficiency at this load.

3.23 A 400-V series motor has a field resistance of 0.2 Ω and an armature resistance of 0.1 Ω. The motor takes 30 A of current at 1000 rpm while developing a torque T. Determine the motor speed if the developed torque is $0.6\,T$.

3.24 A shunt machine, while running as a generator, has an induced voltage of 260 V at 1200 rpm. Its armature and field resistances are 0.2 Ω and 110 Ω, respectively. If the machine is run as a shunt motor, it takes 4 A at 220 V. At a certain load the motor takes 30 A at 220 V. However, on load, armature reaction weakens the field by 3 percent. Calculate the motor speed and efficiency at the specified load.

3.25 The machine of Problem 3.24 is run as a motor. It takes 25 A of current at 800 rpm. What resistance must be inserted in the field circuit to increase the motor speed to 1000 rpm? The torque on the motor for the two speeds remains unchanged.

3.26 The motor of Problem 3.24 runs at 600 rpm while taking 40 A at a certain load. If a 0.8-Ω resistance is inserted in the armature circuit, determine the motor speed provided that the torque on the motor remains constant.

3.27 A 220-V motor delivers 40 hp on full-load at 950 rpm, and has an efficiency of 88 percent. The armature and field resistances are 0.2 Ω and 110 Ω, respectively. Determine (a) the starting resistance such that the starting line current does not exceed 1.6 times the full-load current, and (b) the starting torque.

3.28 A 220-V series motor runs at 750 rpm while taking 15 A of current. What is the motor speed if it takes 10 A of current? The torque on the motor is such that it increases as the square of the speed.

CHAPTER 4

Synchronous Machines

4.1 INTRODUCTION

The bulk of electric power for everyday use is produced by polyphase synchronous generators, which are the largest single-unit electric machines in production. For instance, synchronous generators with power ratings of several hundred megavolt-amperes (MVA) are fairly common, and it is expected that machines of several thousand megavolt-amperes will be in use in the near future. These are called synchronous machines because they operate at constant speeds and constant frequencies under steady-state conditions. Like most rotating machines, synchronous machines are capable of operating both as a motor and as a generator. They are used as motors in constant-speed drives, and where a variable-speed drive is required, a synchronous motor is used with an appropriate frequency changer such as an inverter or cycloconverter. As generators, several synchronous machines often operate in parallel, as in a power station. While operating in parallel, the generators share the load with each other; at a given time one of the generators may not carry any load. In such a case, instead of shutting down the generator, it is allowed to "float" on the line as a synchronous motor on no-load.

The operation of a synchronous generator is based on Faraday's law of electromagnetic induction, and an ac synchronous generator works very much like a dc generator, in which the generation of emf is by the relative motion of conductors and magnetic flux. Clearly, however, a synchronous generator does not have a commutator as does a dc generator. A synchronous generator in its elementary form was shown in Fig. 3.4. The two basic parts of a synchronous machine are the magnetic field structure, carrying a dc-excited winding, and the armature. The armature often has a three-phase winding in which the ac emf is generated. Almost all modern synchronous machines have stationary armatures and rotating field structures. The dc winding on the rotating field structure is connected to an external source through slip rings and brushes. Some field structures do not have brushes but, instead, have brushless excitation by rotating diodes. In some respects the stator carrying the armature windings is similar to the stator of a polyphase induction motor (discussed in Chapter 5).

4.2

SOME CONSTRUCTION DETAILS

Some of the factors that dictate the form of construction of a synchronous machine are discussed next.

Form of Excitation. Notice from the preceding remarks that the field structure is usually the rotating member of a synchronous machine and is supplied with a dc-excited winding to produce the magnetic flux. This dc excitation may be provided by a self-excited dc generator mounted on the same shaft as the rotor of the synchronous machine. Such a generator is known as an *exciter*. The direct current thus generated is fed to the synchronous machine field winding. In slow-speed machines with large ratings, such as hydroelectric generators, the exciter may not be self-excited. Instead, a pilot exciter, which may be self-excited or may have a permanent magnet, activates the exciter. A hydroelectric generator, its rotor, and exciters are shown in Fig. 4.1. The maintenance problems of direct-coupled dc generators impose a limit on this form of excitation at about a 100-MW rating.

An alternative form of excitation is provided by silicon diodes and thyristors, which do not present excitation problems for large synchronous machines. The two types of solid-state excitation systems are:

1. Static systems that have stationary diodes or thyristors, in which the current is fed to the rotor through slip rings.
2. Brushless systems that have shaft-mounted rectifiers that rotate with the rotor, thus avoiding the need for brushes and slip rings. Figure 4.2 shows a brushless excitation system.

FIGURE 4-1. A hydroelectric generator (Courtesy of Brown Boveri Company).

Field Structure and Speed of Machine. We have already mentioned that the synchronous machine is a constant-speed machine. This speed is known as synchronous speed. For instance a 60-Hz two-pole synchronous machine must run at 3600 rpm, whereas the synchronous speed of a 12-pole 60-Hz machine is only 600 rpm. The rotor field structure consequently depends on the speed rating of the machine. Therefore, turbogenerators, which are high-speed machines, have *round* or *cylindrical rotors* (see Figs. 4.3 and 4.4). Hydroelectric

FIGURE 4-2. Rotor of a 3360 kVA 6kV brushless synchronous generator, with rotating diodes (Courtesy of Brown Boveri Company).

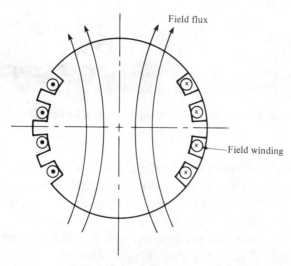

FIGURE 4-3. Field winding on a round or cylindrical rotor.

FIGURE 4-4. Turbine rotor with direct water cooling during the mounting of damper hollow bars (Courtesy of Brown Boveri Company).

and diesel-electric generators are low-speed machines and have *salient pole rotors*, as depicted in Figs. 4.1, 4.2, and 4.5. Such rotors are less expensive than round rotors to fabricate. They are not suitable for large high-speed machines, however, because of the excessive centrifugal forces and mechanical stresses that develop at speeds around 3600 rpm.

Stator. The stator of a synchronous machine carries the armature or load winding. We recall from Chapter 3 that the armature of a dc machine has a winding that is distributed around the periphery of the armature. Thus slot-embedded conductors, covering the entire surface of the armature and interconnected in a predetermined manner, constitute the armature winding of a dc machine. Similarly, in a synchronous machine, the armature winding is formed by interconnecting the various conductors in the slots spread over the periphery of the stator of the machine. Often, more than one independent winding is on the stator. An arrangement of a three-phase stator winding is shown in Fig. 4.6. Notice that the three phases are displaced from each other in space, as the windings are distributed in the slots over the entire periphery of the stator. Each slot contains two coil sides. For instance, slot 1 has coil sides of phases A and B, whereas slot 2 contains two layers (or two coil sides) of phase A only. Such a winding is known as a *double-layer* winding. Furthermore, it is a four-pole winding laid in 36 slots and we thus have three slots per pole per phase.

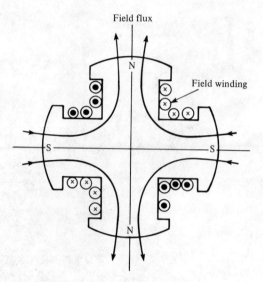

FIGURE 4-5. Field winding on a salient rotor.

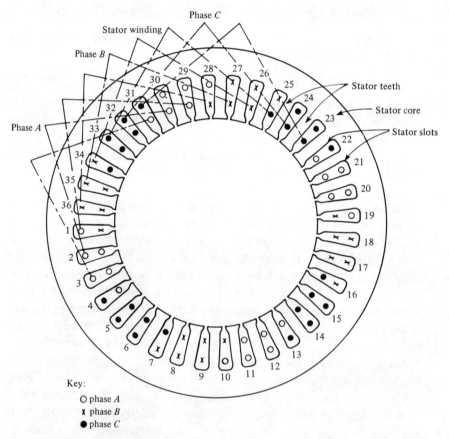

Key:
○ phase A
× phase B
● phase C

FIGURE 4-6. Stator windings.

In order to produce the four-pole flux, each coil should have a span (or *pitch*) of one-fourth of the periphery. In practice, the pitch is made a little less and, as shown in Fig. 4.6, each coil embraces eight teeth. The coil pitch is about 89 percent of the pole pitch, and the winding is, therefore, a *fractional-pitch* (or *chorded*) *winding*. There is essentially no difference between the stator of a round-rotor machine and that of a salient-rotor machine. The stators of waterwheel or hydroelectric generators, however, usually have a large-diameter armature compared to other types of generators. The stator core consists of punchings

FIGURE 4-7. Mounting stator conductors in slots of one stator half of a synchronous motor (Courtesy of Brown Boveri Company).

4.2 SOME CONSTRUCTION DETAILS

of high-quality laminations having slot-embedded windings, as shown in Fig. 4.6. Mounting of stator conductors in slots of one stator half to make the armature winding of a large synchronous machine is shown in Fig. 4.7.

Cooling. Because synchronous machines are often built in extremely large sizes, they are designed to carry very large currents. A typical armature current density may be of the order of 10 A/mm^2 in a well-designed machine. Also, the magnetic loading of the core is such that it reaches saturation in many regions. The severe electric and magnetic loadings in a synchronous machine produce heat that must be appropriately dissipated. Thus the manner in which the active parts of a machine are cooled determines its overall physical structures. In addition to air, some of the coolants used in synchronous machines include water, hydrogen, and helium.

Damper Bars. So far we have mentioned only two electrical windings of a synchronous machine: the three-phase armature winding and the field winding. We also pointed out that, under steady state, the machine runs at a constant speed, that is, at synchronous speed. However, like other electric machines, a synchronous machine undergoes transients during starting and abnormal conditions. During transients, the rotor may undergo mechanical oscillations and its speed deviates from the synchronous speed, which is an undesirable phenomenon. To overcome this, an additional set of windings, resembling the cage of an induction motor, is mounted on the rotor. When the rotor speed is different

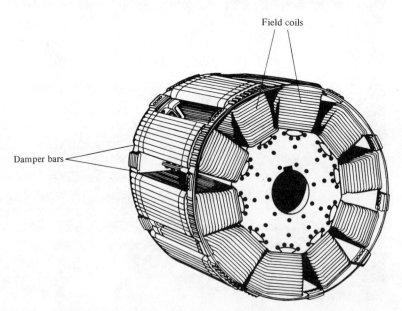

FIGURE 4-8. A salient rotor showing the field windings and damper bars (shaft not shown).

from the synchronous speed, currents are induced in the damper windings. The damper winding acts like the cage rotor of an induction motor, producing a torque to restore the synchronous speed. Also, the damper bars provide a means of starting the machine, which is otherwise not self-starting. Figures 4.4 and 4.8 show the damper bars on round and salient rotors, respectively.

4.3

MAGNETOMOTIVE FORCES AND FLUXES DUE TO ARMATURE AND FIELD WINDINGS

In general, we may say that the behavior of an electric machine depends on the interaction between the magnetic fields (or fluxes) produced by various mmf's acting on the magnetic circuit of the machines. For instance, in a dc motor the torque is produced by the interaction of the flux produced by the field winding and the flux produced by the current-carrying armature conductors. In a dc generator also, we must consider the effect of the interaction between the field and armature mmf's (as discussed in Section 3.3).

As mentioned in the preceding section, the main sources of fluxes in a synchronous machine are the armature and the field mmf's. In contrast with a dc machine, in which the flux due to the armature mmf is stationary in space, the fluxes due to each phase of the armature mmf pulsate in time and the resultant flux rotates in space, as will be demonstrated later. For the present we shall consider the mmf's produced by a single full-pitch coil having N turns, as shown in Fig. 4.9, where the slot opening is negligible. Clearly, the machine has two poles [Fig. 4.9(a)]. The mmf has a constant value of Ni between the coil sides, as shown in Fig. 4.9(b). Traditionally, the magnetic effects of a winding in an electric machine are considered on a per pole basis. Thus if i is the current in the coil, the mmf per pole is $Ni/2$, which is plotted in Fig. 4.9(c). The reason for such a representation is that Fig. 4.9(c) also represents a flux-density distribution, but to a different scale. Obviously, the flux density over one pole (say the north pole) must be opposite to that over the other (south) pole, thus keeping the flux entering the rotor equal to that leaving the rotor surface. Comparing Fig. 4.9(b) and (c), we notice that the representation of the mmf curve with positive and negative areas [Fig. 4.9(c)] has the advantage that it gives the flux-density distribution, which must contain positive and negative areas. The mmf distribution shown in Fig. 4.9(c) may be resolved into its harmonic components. The mmf of one phase of the winding shown in Fig. 4.6 can be obtained by appropriately adding the mmf's of individual coils (see Examples 4.2 and 4.3). In such properly designed armature windings, we may assume harmonics to be absent and the resultant mmf of each phase may be ideally taken as sinusoidal. Such an assumption considerably simplifies the mathematical analysis. A sinu-

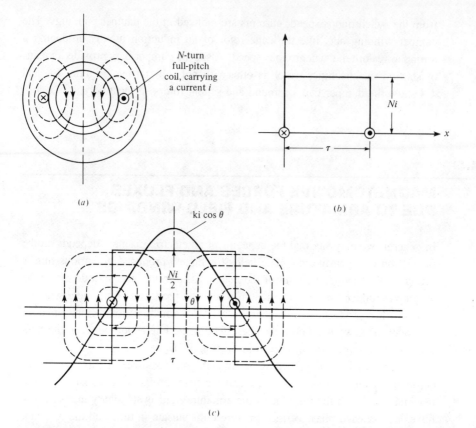

FIGURE 4-9. Flux and mmf produced by a concentrated winding: (a) flux lines produced by an *N*-turn coil; (b) mmf produced by the *N*-turn coil; (c) mmf per pole.

soidally distributed mmf (or the fundamental component of the mmf $Ni/2$) is shown in Fig. 4.9(c). This mmf is mathematically expressed as

$$\mathcal{F} = ki \cos \theta \tag{4.1}$$

where k is a constant, i the current in the winding producing the mmf, and θ the angle measured with respect to the magnetic axis of the winding, as shown in Fig. 4.9(c). Notice from (4.1) that in one complete rotation in the air gap, corresponding to 360°, we obtain one complete cycle of the mmf. One cycle of mmf is said to correspond to 360 electrical degrees (in contrast to the 360 mechanical degrees in one rotation). It is possible to obtain more than one complete cycle of mmf in one complete rotation. For instance, in Fig. 4.10, the winding around the stator periphery is so arranged that we obtain four cycles of mmf. In other words, we obtain 1440 electrical degrees by going through 360 mechanical degrees. The "cycles of mmf's" are precisely designated by the

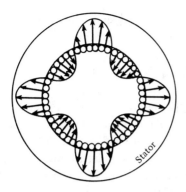

FIGURE 4-10. Four cycles of mmf correspond to 8 poles or 1440° electrical degrees.

number of poles, P. *One cycle of mmf corresponds to two poles and 360 electrical degrees,* and one complete rotation around the machine periphery corresponds to 360 mechanical degrees. Hence for a *P-pole* machine, the general relationship between the electrical degree θ and the mechanical degree θ_m is

$$\theta = \frac{P}{2} \theta_m \qquad (4.2)$$

Obviously, $P/2$ is the number of *pole pairs* and is designated by p. Now we can generalize (4.1) for an mmf having p-pole pairs as

$$\mathcal{F} = ki \cos p\theta \qquad (4.3)$$

Finally, if the winding current $i = I_m \sin \omega t$, then (4.3) becomes

$$\mathcal{F} = \mathcal{F}_m \sin \omega t \cos p\theta \qquad (4.4)$$

where $\mathcal{F}_m = kI_m$ = amplitude of mmf wave.

Now let us consider the three-phase stator (or armature) winding of a synchronous machine to be excited by three-phase currents. As a result, the mmf's produced by the three phases are displaced from each other by 120° in time and space. (Recall from Fig. 4.6 that the armature windings of the three phases are displaced from each other in space.) If we assume the mmf distribution in space to be sinusoidal, we may write for the three mmf's:

$$\mathcal{F}_a = \mathcal{F}_m \sin \omega t \cos p\theta$$

$$\mathcal{F}_b = \mathcal{F}_m \sin (\omega t - 120°) \cos (p\theta - 120°)$$

$$\mathcal{F}_c = \mathcal{F}_m \sin (\omega t + 120°) \cos (p\theta + 120°)$$

4.3 FORCES AND FLUXES

Observing that $\sin A \cos B = \frac{1}{2} \sin (A - B) + \frac{1}{2} \sin (A + B)$ and adding $\mathscr{F}_a$, $\mathscr{F}_b$, and $\mathscr{F}_c$, we obtain the resultant mmf as

$$\mathscr{F}(\theta, t) = 1.5 \mathscr{F}_m \sin (\omega t - p\theta) \tag{4.5}$$

The magnetic field resulting from the mmf of (4.5) is a rotating magnetic field. Graphically, the production of the rotating field is illustrated in Fig. 4.11. Figure 4.12 shows the position of the resultant mmf at three different instants $t_1 < t_2 < t_3$. Notice that as time elapses, a fixed point P moves to the right, implying that the resultant mmf is a traveling wave of a constant amplitude. The magnetic field produced by this mmf in an electric machine is then known as a *rotating magnetic field*. We may arrive at the same conclusion by considering the resultant mmf at various instants, as shown in Fig. 4.11. From these diagrams it is clear that as we progress in time from t_1 to t_3, the resultant mmf rotates in space from θ_1 to θ_3. The existence of the rotating magnetic field is essential to the operation of a synchronous motor.

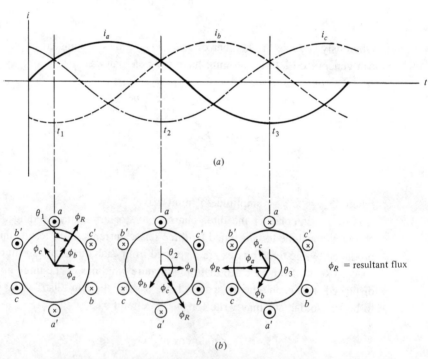

FIGURE 4-11. Production of a rotating magnetic field by a three phase excitation. (a) Time diagram. (b) Space diagram.

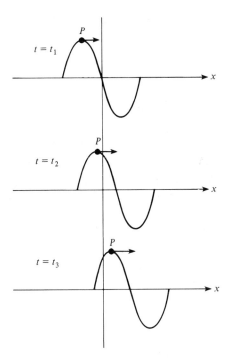

FIGURE 4-12. The function sin $[\omega t - p\theta]$ at different time intervals $t_1 < t_2 < t_3$.

4.4

SYNCHRONOUS SPEED

To determine the velocity of the traveling field given by (4.5), imagine an observer traveling with the mmf wave from a certain point. To this observer, the magnitude of the mmf wave will remain constant (independent of time), implying that the right side of (4.1) would appear constant. Expressed mathematically, this would mean that

$$\sin(\omega t - p\theta) = \text{constant}$$

or

$$\omega t - p\theta = \text{constant}$$

Differentiating both sides with respect to t, we obtain

$$\omega - p\dot{\theta} = 0$$

or

$$\omega_m \equiv \dot{\theta} = \frac{\omega}{p} \quad (4.6)$$

This speed is known as the *synchronous speed*, which is the angular velocity of the mmf wave.

In (4.5) and (4.6), ω is the frequency of the stator mmf's. What is the significance of p? Notice that the given mmf's vary in space as $\cos p\theta$, indicating that for one complete travel around the stator periphery, the mmf undergoes p cyclic changes. Thus p may be considered as the order of harmonics, or the number of *pole pairs* in the mmf wave. If P is the *number of poles* then, obviously, $P = 2p$. Writing ω_m in terms of speed in rpm, n_s, and ω in terms of the frequency, f, we have

$$\omega_m = \frac{2\pi n_s}{60}$$

and

$$\omega = 2\pi f$$

Substituting for p, ω_m, and ω in (4.2) yields

$$n_s = \frac{120f}{P} \quad (4.7)$$

which is the *synchronous speed* in rpm.

An alternative form of (4.7) is

$$f = \frac{Pn_s}{120} \quad (4.8)$$

which implies that the frequency of the voltage induced in a synchronous generator having P poles and running at n_s rpm is f hertz. The same conclusion may be arrived at from Fig. 3.4: namely, in a two-pole machine, one cycle is generated in one rotation. Thus, in a P-pole machine, $P/2$ cycles are generated in one rotation; and in n_s rotations, $Pn_s/2$ cycles are generated. Since n_s rotations take 60 s, in 1 s $Pn_s/2 \times 60 = Pn_s/120$ cycles are generated, which is the frequency f.

EXAMPLE 4.1
For a 60-Hz generator, list four possible combinations of number of poles and speed.

SOLUTION
From (4.8), we must have $Pn_s = 7200 = 120 \times 60$. Hence we obtain the following table:

Number of Poles	Speed (rpm)
2	3600
4	1800
6	1200
8	900

■

EXAMPLE 4.2
An N-turn winding is made up of coils distributed in slots, such as the winding shown in Fig. 4.6. The voltages induced in these coils are displaced from one another in phase by the slot angle α. The resultant voltage at the terminals of the N-turn winding is then the phasor sum of the coil voltages. Find an expression for the *distribution factor*, k_d, where

$$k_d \equiv \frac{\text{magnitude of resultant voltage}}{\text{sum of magnitudes of individual coil voltages}} \qquad (4.9)$$

SOLUTION
Let P be the number of poles, Q the number of slots, and m the number of phases. Then $Q = Pqm$, where q is the number of slots per pole per phase. The slot angle α is given (in electrical degrees) by

$$\alpha = \frac{(180°)P}{Q} = \frac{180°}{mq}$$

The phasor addition of voltages (for $q = 3$) is shown in Fig. 4.13, from the geometry of which we get

$$k_d = \frac{E_r}{qE_c} = \frac{2l \sin(q\alpha/2)}{q[2l \sin(\alpha/2)]} = \frac{\sin(q\alpha/2)}{q \sin(\alpha/2)} \qquad (4.10)$$

which is the desired result. In (4.10), l is the radius of the arc of the circle, as shown in Fig. 4.13. ■

The distribution factors for a few three-phase windings are as follows:

Slot/Pole/Phase	2	3	4	5	6	8
k_d	0.966	0.960	0.958	0.957	0.957	0.956

4.4 SYNCHRONOUS SPEED

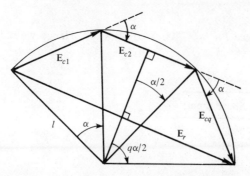

FIGURE 4-13. Phasor addition of individual coil voltages to obtain the distribution factor.

EXAMPLE 4.3

The voltage induced in a fractional-pitch coil is reduced by a factor known as the *pitch factor*, k_p, as compared to the voltage induced in a full-pitch coil. Derive an expression for the pitch factor. Recall that the winding of Fig. 4.6 is a fractional-pitch winding.

SOLUTION

In a sinusoidally distributed flux density we show a full-pitch and a fractional-pitch coil in Fig. 4.14. The coil span of the full-pitch coil is equal to the pole pitch, τ. Let the coil span of the fractional-pitch coil be $\beta < \tau$, as shown. The flux linking the fractional-pitch coil will be proportional to the shaded area in Fig. 4.14, whereas the flux linking the full-pitch coil is proportional to the entire area under the curve. The pitch factor is therefore the ratio of the shaded area to the total area:

$$k_p = \int_{(\tau-\beta)/2}^{(\tau+\beta)/2} \sin \frac{\pi x}{\tau} \, dx \bigg/ \int_0^{\tau} \sin \frac{\pi x}{\tau} \, dx = \sin \frac{\pi \beta}{2\tau} \qquad (4.11)$$

Notice that in (4.11), β and τ may be measured in any convenient unit and $\pi\beta/2\tau$ is in electrical radians. ∎

EXAMPLE 4.4

Sometimes it is convenient to combine the distribution factor and the pitch factor as one factor, called the *winding factor*, k_w. Calculate the distribution factor, the pitch factor, and the winding factor, $k_w \equiv k_d k_p$, for the stator winding of Fig. 4.6.

SOLUTION

From Fig. 4.6, $m = 3$, $P = 4$, and $Q = 36$. Thus

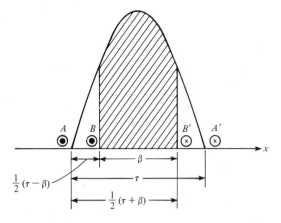

FIGURE 4-14. Determination of pitch factor.

$$q = \frac{36}{(4)(3)} = 3 \qquad \alpha = \frac{180°}{(3)(3)} = 20°$$

Substituting these in (4.10) yields

$$k_d = \frac{\sin 30°}{3 \sin 10°} = 0.96$$

Also, Fig. 4.6 shows that $\tau = 9$ slots and $\beta = 8$ slots. Hence, from (4.11),

$$k_p = \sin \frac{8\pi}{18} = \sin 80° = 0.985$$

and

$$k_w = k_d k_p = (0.96)(0.985) = 0.945$$

4.5

SYNCHRONOUS GENERATOR OPERATION

Like the dc generator, a synchronous generator functions on the basis of Faraday's law. If the flux linking the coil changes in time, a voltage is induced in a coil. Stated in another form, a voltage is induced in a conductor if it cuts magnetic flux lines (see Fig. 3.4). Considering the machine shown in Fig. 4.15 and assuming that the flux density in the air gap is uniform implies that sinusoidally varying voltages will be induced in the three coils aa', bb', and cc' if

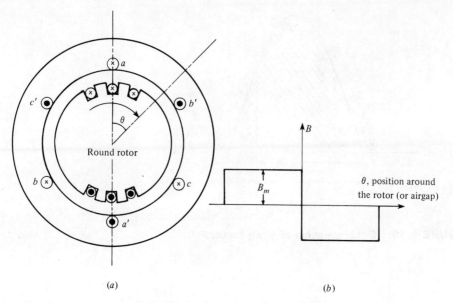

FIGURE 4-15. (a) A 3-phase round rotor machine; (b) flux density distribution produced by the rotor excitation.

the rotor, carrying dc, rotates at a constant speed, n_s. Recalling from Chapter 3 that if φ is the flux per pole, ω is the angular frequency, and N is the number of turns in phase a (coil aa'), then the voltage induced in phase a is given by, according to (3.3),

$$e_a = \omega N \varphi \sin \omega t \tag{4.12}$$
$$= E_m \sin \omega t$$

where $E_m = 2\pi f N \varphi$ and $f = \omega/2\pi$ is the frequency of the induced voltage. Because phases b and c are displaced from phase a by $\pm 120°$ (Fig. 4.15), the corresponding voltages may be written as

$$e_b = E_m \sin(\omega t - 120°)$$
$$e_c = E_m \sin(\omega t + 120°)$$

These voltages are sketched in Fig. 4.16 and correspond to the voltages from a three-phase generator.

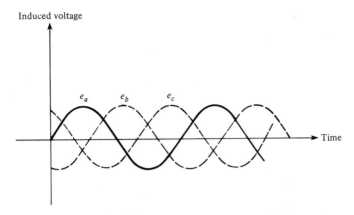

FIGURE 4-16. A three-phase voltage produced by a three-phase synchronous generator.

4.6

PERFORMANCE OF A ROUND-ROTOR SYNCHRONOUS GENERATOR

At the outset we wish to point out that we will study the machine on a per phase basis, implying a balanced operation. Thus let us consider a round-rotor machine operating as a generator on no-load. Variation of E_0 with I_f is shown in Fig. 4.17 and is known as the open-circuit characteristic of a synchronous generator. Let the open-circuit phase voltage be E_0 for a certain field current I_f. Here E_0 is the internal voltage of the generator. We assume that I_f is such that the machine is operating under unsaturated condition. Next we short-circuit the armature at the terminals, keeping the field current unchanged (at I_f), and measure the armature phase current I_a. In this case, the entire internal voltage E is dropped across the internal impedance of the machine. In mathematical terms,

$$\mathbf{E} = \mathbf{I}_a \mathbf{Z}_s$$

and $\mathbf{Z}_s$ is known as the *synchronous impedance*. One portion of $\mathbf{Z}_s$ is R_a and the other a reactance, X_s, known as synchronous reactance; that is,

$$\mathbf{Z}_s = R_a + jX_s \qquad (4.13)$$

If the generator operates at a terminal voltage V_t while supplying a load corresponding to an armature current I_a, then

$$\mathbf{E} = \mathbf{V}_t + \mathbf{I}_a (R_a + jX_s) \qquad (4.14)$$

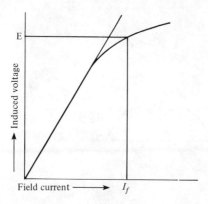

FIGURE 4-17. Open-circuit characteristics of a synchronous generator.

where X_s, the *synchronous reactance*, which exists by virtue of the current-carrying armature windings.

In an actual synchronous machine, except in very small ones, we almost always have $X_s \gg R_a$, in which case $Z_s \simeq jX_s$. We will use this restriction in most of the analysis. Among the steady-state characteristics of a synchronous generator, its voltage regulation and power-angle characteristics are the most important ones. As for a transformer and a dc generator, we define the voltage regulation of a synchronous generator at a given load as

$$\text{percent voltage regulation} = \frac{E - V_t}{V_t} \times 100 \qquad (4.15)$$

where V_t is the terminal voltage on load and E is the no-load terminal voltage. Clearly, for a given V_t, we can find E from (4.14) and hence the voltage regulation, as illustrated by the following examples.

EXAMPLE 4.5

Calculate the percent voltage regulation for a three-phase wye-connected 2500-kVA 6600-V turboalternator operating at full-load and 0.8 power factor lagging. The per phase synchronous reactance and the armature resistance are 10.4 Ω and 0.071 Ω, respectively.

SOLUTION

Clearly, we have $X_s \gg r_a$. The phasor diagram for the lagging power factor, neglecting the effect or r_a, is shown in Fig. 4.18(a). The numerical values are as follows:

$$V_t = \frac{6600}{\sqrt{3}} = 3810 \text{ V}$$

$$I_a = \frac{2500 \times 1000}{\sqrt{3} \times 6600} = 218.7 \text{ A}$$

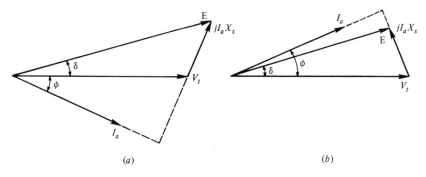

FIGURE 4-18. Phasor diagrams: (a) lagging power factor; (b) leading power factor.

From (4.14) we have

$$\mathbf{E} = 3810 + 218.7(0.8 - j0.6)j10.4 = 5485\,\underline{/19.3°}$$

and, from (4.15),

$$\text{percent regulation} = \frac{5485 - 3810}{3810} \times 100 = 44\%$$ ∎

EXAMPLE 4.6
Repeat the preceding calculations with 0.8 power factor leading.

SOLUTION
In this case we have the phasor diagram shown in Fig. 4.18(b), from which we get

$$\mathbf{E} = 3810 + 218.7(0.8 + j0.6)j10.4 = 3048\,\underline{/36.6°}$$

and

$$\text{percent voltage regulation} = \frac{3048 - 3810}{3810} \times 100 = -20\%$$ ∎

We observe from these examples that the voltage regulation is dependent on the power factor of the load. Unlike what happens in a dc generator, the voltage regulation for a synchronous generator may even become negative. The angle between $\mathbf{E}$ and $\mathbf{V}_t$ is defined as the *power angle*, δ. Notice that the power angle, δ, is not the same as the power factor angle, φ. To justify this definition, we consider Fig. 4.18(a), from which we obtain

$$I_a X_s \cos \varphi = E \sin \delta \tag{4.16}$$

4.6 SYNCHRONOUS GENERATOR PERFORMANCE

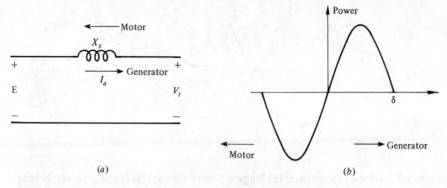

FIGURE 4-19. (a) an approximate equivalent circuit; (b) power–angle characteristics of a round-rotor synchronous machine.

Now, from the approximate equivalent circuit (assuming that $X_s \gg R_a$), the power delivered by the generator = power developed, $P_d = V_t I_a \cos\varphi$ [which follows from Fig. 4.18(a) also]. Hence, in conjunction with (4.16), we get

$$P_d = \frac{E V_t}{X_s} \sin\delta \qquad (4.17)$$

which shows that the internal power of the machine is proportional to $\sin\delta$. Equation (4.17) is often said to represent the *power-angle characteristic* of a synchronous machine. A plot of (4.17) is shown in Fig. 4.19(b), which shows that for a negative δ, the machine will operate as a motor, as discussed in the next section.

4.7

SYNCHRONOUS MOTOR OPERATION

We know (from Section 4.3) that the stator of a three-phase synchronous machine, carrying a three-phase excitation, produces a rotating magnetic field in the air gap of the machine. Referring to Fig. 4.20, we will have a rotating magnetic field in the air gap of the salient pole machine when its stator (or armature) windings are fed from a three-phase source. Let the rotor (or field) winding be unexcited. The rotor will have a tendency to align with the rotating field at all times in order to present the path of least reluctance. Thus if the field is rotating, the rotor will tend to rotate with the field. From Fig. 4.15(a), we see that a round rotor will not tend to follow the rotating magnetic field because the uniform air gap presents the same reluctance all around the air gap and the rotor does not have any preferred direction of alignment with the magnetic field. This torque, which we have in the machine of Fig. 4.20 but not in the machine

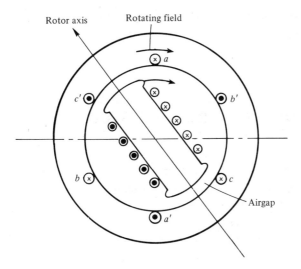

FIGURE 4-20. A salient-rotor machine.

of Fig. 4.15(a), is called the *reluctance torque*. It is present by virtue of the variation of the reluctance around the periphery of the machine.

Next, let the field winding [Fig. 4.15(a) or 4.20] be fed by a dc source that produces the rotor magnetic field of definite polarities, and the rotor will tend to align with the stator field and will tend to rotate with the rotating magnetic field. We observe that for an excited rotor, a round rotor, or a salient rotor, both will tend to rotate with the rotating magnetic field, although the salient rotor will have an additional reluctance torque because of the saliency. In a later section we derive expressions for the electromagnetic torque in a synchronous machine attributable to field excitation and to saliency.

So far we have indicated the mechanism of torque production in a round-rotor and in a salient-rotor machine. To recapitulate, we might say that the stator rotating magnetic field has a tendency to "drag" the rotor along, as if a north pole on the stator "locks in" with a south pole of the rotor. However, if the rotor is at a standstill, the stator poles will tend to make the rotor rotate in one direction and then in the other as they rapidly rotate and sweep across the rotor poles. Therefore, a synchronous motor is not self-starting. In practice, as mentioned earlier, the rotor carries damper bars that act like the cage of an induction motor and thereby provide a starting torque. The mechanism of torque production by the damper bars is similar to the torque production in an induction motor, discussed in Chapter 5. Once the rotor starts running and almost reaches the synchronous speed, it locks into position with the stator poles. The rotor pulls into step with the rotating magnetic field and runs at the synchronous speed; the damper bars go out of action. Any departure from the synchronous speed results in induced currents in the damper bars, which tend to restore the synchronous speed. Machines without damper bars, or very large machines with damper bars, may be started by an auxiliary motor. We discuss the operating characteristics of synchronous motors in the next section.

4.8 PERFORMANCE OF A ROUND-ROTOR SYNCHRONOUS MOTOR

Except for some precise calculations, we may neglect the armature resistance as compared to the synchronous reactance. Therefore, the steady-state per phase equivalent circuit of a synchronous machine simplifies to the one shown in Fig. 4.19(a). Notice that this circuit is similar to that of a dc machine, where the dc armature resistance has been replaced by the synchronous reactance. In Fig. 4.19(a) we have shown the terminal voltage V_t, the internal excitation voltage E, and the armature current I_a going "into" the machine or "out of" it, depending on the mode of operation—"into" for motor and "out of" for generator. With the help of this circuit and (4.17) we will study some of the steady-state operating characteristics of a synchronous motor. In Fig. 4.19(b) we show the power-angle characteristics as given by (4.17). Here positive power and positive δ imply the generator operation, while a negative δ corresponds to a motor operation. Because δ is the angle between $\mathbf{E}$ and $\mathbf{V}_t$, $\mathbf{E}$ is ahead of $\mathbf{V}_t$ in a generator, whereas in a motor, $\mathbf{V}_t$ is ahead of $\mathbf{E}$. The voltage-balance equation for a motor is, from Fig. 4.19(a),

$$\mathbf{V}_t = \mathbf{E} + j\mathbf{I}_a\mathbf{X}_s$$

If the motor operates at a constant power, then (4.16) and (4.17) require that

$$E \sin \delta = I_a X_s \cos \varphi = \text{constant} \quad (4.18)$$

We recall that E depends on the field current, I_f (see Fig. 4.17). Consider two cases: (1) when I_f is adjusted so that $E < V_t$ and the machine is underexcited, and (2) when I_f is increased to a point that $E > V_t$ and the machine becomes overexcited. The voltage–current relationships for the two cases are shown in Fig. 4.21(a). For $E > V_t$ at constant power, δ is greater than the δ for $E < V_t$, as governed by (4.14). Notice that an underexcited motor operates at a lagging power factor (I_a lagging V_t), whereas an overexcited motor operates at a leading power factor. In both cases the terminal voltage and the load on the motor are the same. Thus we observe that the operating power factor of the motor is controlled by varying the field excitation, hence altering E. This is a very important property of synchronous motors. The locus of the armature current at a constant load, as given by (4.18), for varying field current is also shown in Fig. 4.21(a). From this we can obtain the variations of the armature current I_a with

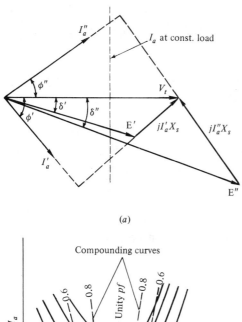

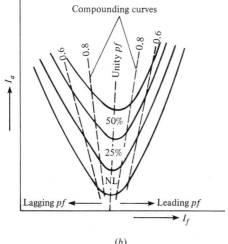

FIGURE 4-21. (a) Phasor diagram for motor operation (E', I_a', φ', and δ') correspond to underexcited operation. (E", I_a'', φ'', and δ'') correspond to overexcited operation. (b) V-curves of a synchronous motor.

the field current, I_f (corresponding to E), and this can be done for different loads, as shown in Fig. 4.21(b). These curves are known as the *V curves* of the synchronous motor. One of the applications of a synchronous motor is in power factor correction, as demonstrated by the following examples. In addition to the V curves, we have also shown the curves for constant power factors. These curves are known as *compounding curves*.

In the preceding paragraph, we have discussed the effect of change in the field current on the synchronous machine power factor. However, the load supplied by a synchronous machine cannot be varied by changing the power factor. Rather, the load on the machine is varied by instantaneously changing the speed (in case of a generator, by supplying additional power by the prime mover), and thus changing the power angle corresponding to the new load. In a synchronous motor, a load change results in a change in the power angle.

EXAMPLE 4.7

A three-phase wye-connected load takes 50 A of current at 0.707 lagging power factor at 220 V between the lines. A three-phase wye-connected, round-rotor synchronous motor, having a synchronous reactance of 1.27 Ω per phase, is connected in parallel with the load. The power developed by the motor is 33 kW at a power angle, δ, of 30°. Neglecting the armature resistance, calculate (a) the reactive kilovolt-amperes (kvar) of the motor and (b) the overall power factor of the motor and the load.

SOLUTION

(a) The circuit and the phasor diagram on a per phase basis are shown in Fig. 4.22. From (4.17) we have

$$P_d = \frac{1}{3} \times 33{,}000 = \frac{220}{\sqrt{3}} \frac{E}{1.27} \sin 30°$$

which yields $E = 220$ V. From the phasor diagram, $I_a X_s = 127$ or $I_a = 127/1.27 = 100$ A and $\varphi_a = 30°$. The reactive kilovolt-amperes of the motor $= \sqrt{3} \times V_t I_a \sin \varphi_a = \sqrt{3} \times 220/1000 \times 100 \times \sin 30 = 19$ kvar.

(b) Notice that φ_a, the power-factor angle of the motor, φ_L, the power-factor angle of the load, and φ, the overall power-factor angle, are shown in Fig. 4.22(b). The power angle, δ, is also shown in this phasor diagram, from which

$$\mathbf{I} = \mathbf{I}_L + \mathbf{I}_a$$

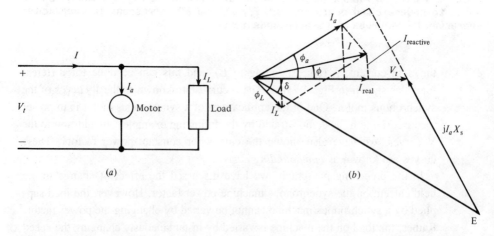

FIGURE 4-22. (a) circuit diagram; (b) phasor diagram.

Or algebraically adding the real and reactive components of the currents, we obtain

$$I_{\text{real}} = I_a \cos \varphi_a + I_L \cos \varphi_L$$

$$I_{\text{reactive}} = I_a \sin \varphi_a - I_L \sin \varphi_L$$

The overall power factor angle, φ, is thus given by

$$\tan \varphi = \frac{I_a \sin \varphi_a - I_L \sin \varphi_L}{I_a \cos \varphi_a + I_L \cos \varphi_L} = 0.122$$

or $\varphi = 7°$ and $\cos \varphi = 0.992$ leading. ∎

EXAMPLE 4.8

For the generator of Example 4.5 calculate the power factor for zero voltage regulation on full load.

SOLUTION

Let φ be the power factor angle. Then

$$\mathbf{I}_a \mathbf{Z}_s = 218.7 \times 10.4 \underline{/\varphi + 89.6} = 2274.48 \underline{/\varphi + 89.6} \text{ V}$$

For voltage regulation to be zero, $|\mathbf{E}| = |\mathbf{V}_t|$, where $\mathbf{E}$ and $\mathbf{V}_t$ are related by

$$\mathbf{E} = \mathbf{V}_t + \mathbf{I}_a \mathbf{Z}_s$$

Rewriting the right-hand side in complex form and substituting the numerical values yields

$$\mathbf{E} = 3810 + j0 + 2274.48 \cos (\varphi + 89.6) + j2274.48 \sin (\varphi + 89.6)$$

For zero voltage regulation, $|\mathbf{E}| = 3810$. Hence

$$3810^2 = [3810 + 2274.48 \cos (\varphi + 89.6)]^2 + [2274.48 \sin (\varphi + 89.6)]^2$$

from which

$$\varphi = 17.76 \quad \text{and} \quad \cos \varphi = 0.95 \text{ leading} \quad ∎$$

4.9

SALIENT-POLE SYNCHRONOUS MACHINES

In the preceding discussion we have analyzed the round-rotor machine and made extensive use of the machine parameter, which we defined as synchronous reactance. Because of saliency, the reactance measured at the terminals of a salient-rotor machine will vary as a function of the rotor position. This is not so in a round-rotor machine. Thus a simple definition of the synchronous reactance for a salient-rotor machine is not immediately forthcoming.

To overcome this difficulty, we use the two-reaction theory proposed by André Blondel. The theory proposes to resolve the given armature mmf's into two mutually perpendicular components, with one located along the axis of the rotor salient pole, known as the direct (or d) axis and with the other in quadrature and known as the quadrature (or q) axis. Correspondingly, we may define the d-axis and q-axis synchronous reactances, X_d and X_q, for a salient-pole synchronous machine. Thus, for generator operation, we draw the phasor diagram of Fig. 4.23. Notice that I_a has been resolved into its d- and q-axis (fictitious) components, I_d and I_q. With the help of this phasor diagram, we obtain

$$I_d = I_a \sin(\delta + \varphi) \qquad I_q = I_a \cos(\delta + \varphi)$$
$$V_t \sin \delta = I_q X_q = I_a X_q \cos(\delta + \varphi)$$

From these we get

$$V_t \sin \delta = I_a X_q \cos \delta \cos \varphi - I_a X_q \sin \delta \sin \varphi$$

or

$$(V_t + I_a X_q \sin \varphi) \sin \delta = I_a X_q \cos \delta \cos \varphi$$

Dividing both sides by $\cos \delta$ and solving for $\tan \delta$ yields

$$\tan \delta = \frac{I_a X_q \cos \varphi}{V_t + I_a X_q \sin \varphi} \tag{4.19}$$

With δ known (in terms of φ), the voltage regulation may be computed from

$$E = V_t \cos \delta + I_d X_d$$

$$\text{percent regulation} = \frac{E - V_t}{V_t} \times 100\%$$

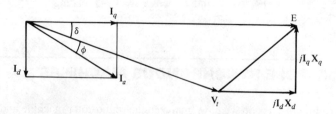

FIGURE 4-23. Phasor diagram of a salient-pole generator.

In fact, the phasor diagram depicts the complete performance characteristics of the machine.

Let us now use Fig. 4.23 to derive the power-angle characteristics of a salient-pole generator. If armature resistance is neglected, $P_d = V_t I_a \cos \varphi$. Now, from Fig. 4.23, the projection of I_a on V_t is

$$\frac{P_d}{V_t} = I_a \cos \varphi = I_q \cos \delta + I_d \sin \delta \tag{4.20}$$

Solving

$$I_q X_q = V_t \sin \delta \quad \text{and} \quad I_d X_d = E - V_t \cos \delta$$

for I_q and I_d, and substituting in (4.13), gives

$$P_d = \frac{EV_t}{X_d} \sin \delta + \frac{V_t^2}{2}\left(\frac{1}{X_q} - \frac{1}{X_d}\right) \sin 2\delta \tag{4.21}$$

Equation (4.21) can also be established for a salient-pole motor ($\delta < 0$); the graph of (4.21) is given in Fig. 4.24. Observe that for $X_d = X_q = X_s$, (4.21) reduces to the round-rotor equation, (4.17).

EXAMPLE 4.9

A 20-kVA 220-V 60-Hz wye-connected three-phase salient-pole synchronous generator supplies rated load at 0.707 lagging power factor. The phase constants of the machine are $R_a = 0.5\ \Omega$ and $X_d = 2X_q = 4.0\ \Omega$. Calculate the voltage regulation at the specified load.

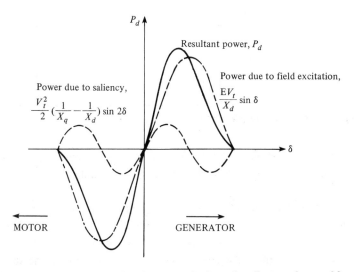

FIGURE 4-24. Power–angle characteristics of salient-pole machines.

SOLUTION

$$V_t = \frac{220}{\sqrt{3}} = 127 \text{ V}$$

$$I_a = \frac{20{,}000}{\sqrt{3} \times 120} = 52.5 \text{ A}$$

$$\varphi = \cos^{-1} 0.707 = 45°$$

From (4.19),

$$\tan \delta = \frac{I_a X_q \cos \varphi}{V_t + I_a X_q \sin \varphi}$$

$$= \frac{52.5 \times 2 \times 0.707}{127 + 52.5 \times 2 \times 0.707} = 0.37$$

or

$$\delta = 20.6°$$

$$I_d = 52.5 \sin(20.6 + 45) = 47.5 \text{ A}$$

$$I_d X_d = 47.5 \times 4 = 190.0 \text{ V}$$

$$E = V_t \cos \delta + I_d X_d$$

$$= 127 \cos 20.6 + 190 = 308 \text{ V}$$

and

$$\text{percent regulation} = \frac{E - V_t}{V_t} \times 100\% = \frac{308 - 127}{127} \times 100 = 142\% \quad \blacksquare$$

4.10

PARALLEL OPERATION

An electric power station often has several synchronous generators operating in parallel with each other. Some of the advantages of parallel operation are:

1. In the absence of one of the several machines, for maintenance or some other reason, the power station can function with the remaining units.
2. Depending on the load, generators may be brought on line, or taken off, and thus result in the most efficient and economical operation of the station.
3. For future expansion, units may be added on and operate in parallel.

In order that a synchronous generator may be connected in parallel with a system (or bus), the following conditions must be fulfilled:

1. The frequency of the incoming generator must be the same as the frequency of the power system to which the generator is to be connected.
2. The magnitude of the voltage of the incoming generator must be the same as the system terminal voltage.
3. With respect to an external circuit, the voltage of the incoming generator must be in the same phase as system voltage at the terminals.
4. In a three-phase system, the generator must have the same phase sequence as that of the bus.

The process of properly connecting a synchronous generator in parallel with a system is known as *synchronizing*. Two generators can be synchronized either by using a synchroscope or lamps. Figure 4.25 shows a circuit diagram showing lamps as well as synchroscope. The potential transformers (PTs) are used to reduce the voltage for instrumentation. Let the generator G_1 be already in operation with its switch S_{g1} closed. Other switches—S_{g2}, S_1, and S_2—are all open.

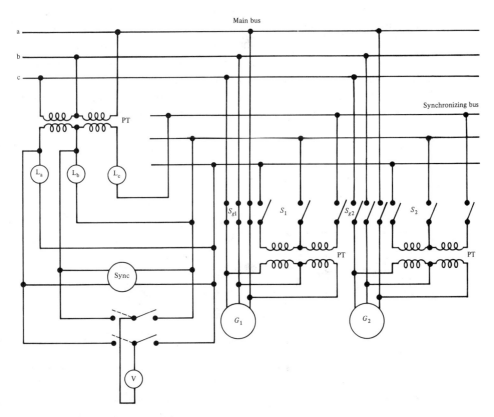

FIGURE 4-25. Synchronizing two generators.

4.10 PARALLEL OPERATION

After the generator G_2 is started and brought up to approximately synchronous speed, S_2 is closed. Subsequently, the lamps—L_a, L_b, and L_c—begin to flicker at a frequency equal to the difference of the frequencies of G_1 and G_2. The equality of the voltages of the two generators is ascertained by the voltmeter V, connected by the double-pole double-throw switch S. Now, if the voltages and frequencies of the two generators are the same, but there is a phase difference between the two voltages, the lamps will glow steadily. The speed of G_2 is then slowly adjusted until the lamps remain permanently dark (because they are connected such that two voltages through them are in opposition). Next, S_{g2} is closed and S_2 may be opened.

In the discussion above, it has been assumed that G_1 and G_2 both have the same phase rotation. On the other hand, let the phase sequence of G_1 be abc counterclockwise and that of G_2 be $a'b'c'$ clockwise. At the synchronous speed of G_1, a and a' may be coincident. This will be indicated by a dark L_a. But L_b and L_c will have equal brightness. When G_2 runs at a speed slightly less than the synchronous speed, the lamps will be dark and bright in the cyclical order L_a, L_b, and L_c. If either of the two conditions prevail, the phase rotation of G_1 must be reversed. This process of testing the phase sequence is known as *phasing out*.

A synchroscope is often used to synchronize two generators which have previously been phased out. A synchroscope is an instrument having a rotating pointer, which indicates whether the incoming machine is slow or fast. One type of synchroscope is shown schematically in Fig. 4.26. It consists of a field coil, F, connected to the main busbars through a large resistance R_F to ensure that the field current is almost in phase with the busbar voltage, V. The rotor consists of two windings R and X, in space quadrature, connected in parallel to

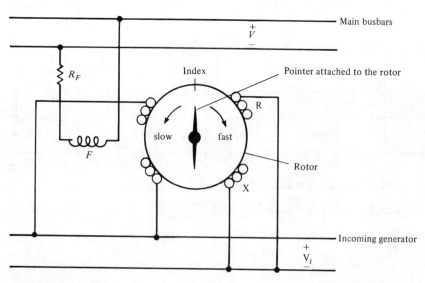

FIGURE 4-26. A synchroscope.

each other and across the incoming generator. The windings R and X are so designed that their respective currents are approximately in phase and 90° behind the terminal voltage, V_i, of the incoming generator. The rotor will align itself so that the axes of R and F are inclined at an angle equal to the phase displacement between V and V_i. If there is a difference between the frequencies of V and V_i, the pointer will rotate at a speed proportional to this difference. The direction of rotation of the pointer will determine if the incoming generator is running below or above synchronism. At synchronism, the pointer will remain stationary at the index.

In present-day power stations, automatic synchronizers are used. But a discussion of these is beyond the scope of this book.

Circulating Current and Load Sharing. At the time of synchronizing (that is, when S_2 of Fig. 4.25 is closed), if G_2 is running at a speed slightly less than that of G_1, the phase relationships of their terminal voltages with respect to the local circuit are as shown in Fig. 4.27(a). The resultant voltage V_c acts in the local circuit to set up a circulating current I_c lagging V_c by a phase angle φ_c. For simplification, if we assume the generators to be identical, then

$$\tan \varphi_c = \frac{R_a}{X_s} \tag{4.22}$$

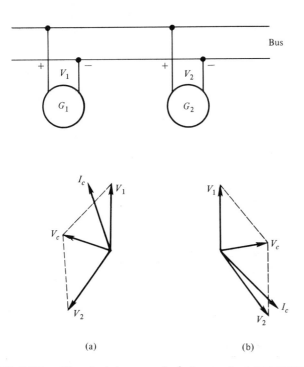

FIGURE 4-27. Circulating currents between two generators.

4.10 PARALLEL OPERATION

and

$$I_c = \frac{V_c}{2Z_s} \quad (4.23)$$

where $R_a + jX_s = \mathbf{Z}_s$ = synchronous impedance R_a = armature resistance and X_s = synchronous reactance.

Notice from Fig. 4.27(a) that $\mathbf{I}_c$ has a component in phase with $\mathbf{V}_1$, and thus acts as a load on G_1 and tends to slow it down. The component of $\mathbf{I}_c$ in phase opposition to $\mathbf{V}_2$ aids G_2 to operate as a motor and thereby G_2 picks up speed. On the other hand, if G_2 was running faster than G_1 at the instant of synchronization, the phase relationships of the voltages and the circulating current become as shown in Fig. 4.27(b). Consequently, G_2 will function as a generator and will tend to slow down; and while acting as a motor, G_1 will pick up speed. Thus there is an inherent synchronizing action which aids the machines to stay in synchronism.

We now recall from Section 4.6 that the power developed by a synchronous machine is given by (4.17). Observe that V_t is the terminal voltage, which is the same as the system busbar voltage. The voltage E is the internal voltage of the generator and is determined by the field excitation. As we have discussed earlier, a change in the field excitation merely controls the power factor and the circulation current at which the synchronous machine operates. The power developed by the machine depends on the power angle δ. For G_2 to share the load, (4.17) must be satisfied. Thus, for a given V_t and E, the power angle must be increased by increasing the prime-mover power. The load sharing between two synchronous generators is illustrated by the following examples.

EXAMPLE 4.10

Two identical three-phase wye-connected synchronous generators share equally a load of 10 MW at 33 kV and 0.8 lagging power factor. The synchronous reactance of each machine is 6 Ω per phase and the armature resistance is negligible. If one of the machines has its field excitation adjusted to carry 125 A of lagging current, what is the current supplied by the second machine? The prime mover inputs to both machines are equal.

SOLUTION

The phasor diagram of current division is shown in Fig. 4.28, wherein $\mathbf{I}_1 = $ 125 A. Because the machines are identical and the prime-mover inputs to both machines are equal, each machine supplies the same true power:

$$I_1 \cos \varphi_1 = I_2 \cos \varphi_2 = \tfrac{1}{2} I \cos \varphi$$

Now

$$I = \frac{10 \times 10^6}{\sqrt{3}(33 \times 10^3)(0.8)} = 218.7 \text{ A}$$

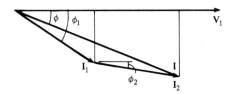

FIGURE 4-28. Example 4.10.

whence

$$I_1 \cos \varphi_1 = I_2 \cos \varphi_2 = \tfrac{1}{2}(218.7)(0.8) = 87.5 \text{ A}$$

The reactive current of the first machine is therefore

$$I_1 |\sin \varphi_1| = \sqrt{(125)^2 - (87.5)^2} = 89.3 \text{ A}$$

and since the total reactive current is

$$I |\sin \varphi| = (218.7)(0.6) = 131.2 \text{ A}$$

the reactive current of the second machine is

$$I_2 |\sin \varphi_2| = 131.2 - 89.3 = 41.9 \text{ A}$$

Hence

$$I_2 = \sqrt{(87.5)^2 + (41.9)^2} = 97 \text{ A} \qquad \blacksquare$$

EXAMPLE 4.11

Consider the two machines of Example 4.10. If the power factor of the first machine is 0.9 lagging and the load is shared equally by the two machines, what are the power factor and current of the second machine?

SOLUTION
Load:

$$\text{power} = 10{,}000 \text{ kW}$$
$$\text{apparent power} = 12{,}500 \text{ kVA}$$
$$\text{reactive power} = -7500 \text{ kvar}$$

First machine:

$$\text{power} = 5000 \text{ kW}$$
$$\varphi_1 = \cos^{-1} 0.9 = -25.8°$$
$$\text{reactive power} = 5000 \tan \varphi_1 = -2422 \text{ kvar}$$

Second machine:

$$\text{power} = 5000 \text{ kW}$$
$$\text{reactive power} = -7500 - (-2422) = -5078 \text{ kvar}$$
$$\tan \varphi_2 = \frac{-5078}{5000} = -1.02$$
$$\cos \varphi_2 = 0.7$$
$$I_2 = \frac{5000}{\sqrt{3}(33)(0.7)} = 124.7 \text{ A}$$

■

4.11 DETERMINATION OF MACHINE CONSTANTS

The synchronous reactance, X_s, of a round-rotor synchronous machine can be obtained from open-circuit and short-circuit tests on the machine. The no-load or open-circuit voltage characteristic of a synchronous generator is similar to that of a dc generator. Figure 4.29 shows such a characteristic, with the effect of magnetic saturation included. Now, if the terminals of the generator are short-circuited, the induced voltage is dropped internally within the generator. The short-circuit current characteristic is also shown in Fig. 4.29. Expressed mathematically (on a per phase basis),

$$\mathbf{E} = \mathbf{I}_a \mathbf{Z}_s = \mathbf{I}_a (R_a + jX_s) \qquad (4.24)$$

In (4.24), E is the no-load armature voltage at a certain field current, and I_a is the short-circuit armature current at the same value of the field current. The impedance Z_s is known as the *synchronous impedance*; R_a is the armature resistance and X_s is defined as the *synchronous reactance*. The synchronous reactance is readily measured for a round-rotor generator, since it is independent of the rotor position in such a machine. In salient-pole generators, however, the synchronous reactance depends on the rotor position.

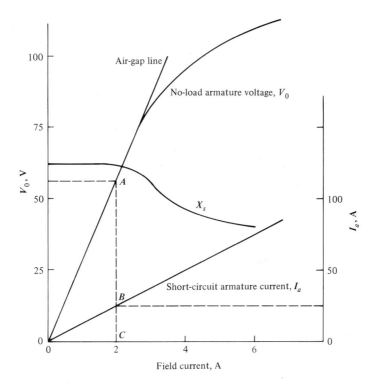

FIGURE 4-29. Open-circuit and short-circuit characteristics of a synchronous generator.

In most synchronous machines, $R_a \ll X_s$, so that in terms of Fig. 4.29,

$$X_s \approx Z_s = \frac{\overline{AC}}{\overline{BC}}$$

Thus X_s varies with field current as indicated by the falling (because of saturation) curve in Fig. 4.29. However, for most calculations, we shall use the linear (constant) value of X_s.

Sudden Short-Circuit at the Armature Terminals. Consider a three-phase generator, on no-load, running at its synchronous speed and carrying a constant field current. Suddenly, the three phases are short-circuited. Symmetrical short-circuit armature current is graphed in Fig. 4.30. Notice that for the first few cycles the current, i_a, decays very rapidly; we term this duration the *subtransient period*. During the next several cycles, the current decreases somewhat slowly, and this range is called the *transient period*. Finally, the current reaches its steady-state value. These currents are limited, respectively, by the *subtransient reactance*, x_d''; the *transient reactance*, x_d'; and the *synchronous reactance*, X_d (or X_s). The subtransient reactance is essentially due to the pres-

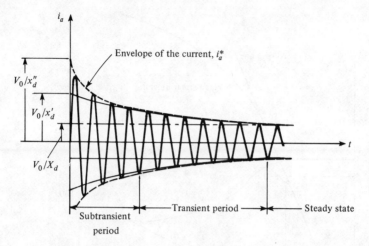

FIGURE 4-30. Armature current under sudden short-circuit.

ence of damper bars, the transient reactance accounts for the field winding, and the synchronous reactance is reactance due to the armature windings. It can be shown that the envelope of the instantaneous armature current (dashed curves in Fig. 4.30) is given by

$$i_a^* = \pm E\left[\left(\frac{1}{x_d''} - \frac{1}{x_d'}\right)e^{-t/\tau_d''} + \left(\frac{1}{x_d'} - \frac{1}{X_d}\right)e^{-t/\tau_d'} + \frac{1}{X_d}\right]$$

where τ_d'' = subtransient time constant
  τ_d' = transient time constant
  E = open-circuit armature phase voltage

The upper branch of the envelope is separately shown in Fig. 4.31.

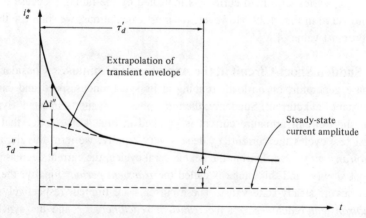

FIGURE 4-31. Envelope of the transient current.

The reactances and the time constants can be determined from design data, but the details are extremely cumbersome. On the other hand, these may be determined from test data and Figs. 4.30 and 4.31, as illustrated by Example 4.12. In Fig. 4.31,

$$\Delta i'' = 0.368(i_d'' - i_d') \qquad \Delta i' = 0.368(i_d' - i_d)$$

where $i_d'' \equiv V_0/x_d''$, $i_d' \equiv V_0/x_d'$, and $i_d \equiv V_0/X_d$. Table 4.1 gives typical values of synchronous machine constants; the per unit values are based on the machine rating.

EXAMPLE 4.12

A three-phase short-circuit test is performed on a synchronous generator for which the envelope of the armature current is shown in Fig. 4.30. Given $E = 231$ V, $\Delta i'' = 113$ A, and $\Delta i' = 117$ A, the steady-state short-circuit current is 144 A. Determine (a) X_d, (b) x_d', and (c) x_d''.

SOLUTION

(a)
$$X_d = \frac{E}{i_d} = \frac{231}{144} = 1.6 \ \Omega$$

(b)
$$\Delta i' = 0.368(i_d' - i_d)$$

$$117 = 0.368(i_d' - 144)$$

$$i_d' = 462 \text{ A}$$

$$x_d' = \frac{E}{i_d'} = \frac{231}{462} = 0.5 \ \Omega$$

TABLE 4.1 Per Unit Synchronous Machine Reactances and Time Constants

Constant	Salient-Pole Machine	Round-Rotor Machine
X_d	1.0–1.25	1.0–1.2
X_q	0.65–0.80	1.0–1.2
x_d'	0.35–0.40	0.15–0.25
x_d''	0.20–0.30	0.10–0.15
τ_d	0.15	0.15
τ_d'	0.9–1.1	1.4–2.0
τ_d''	0.03–0.04	0.03–0.04

(c)
$$\Delta i'' = 0.368(i_d'' - i_d')$$
$$113 = 0.368(i_d'' - 462)$$
$$i_d'' = 769 \text{ A}$$
$$x_d'' = \frac{E}{i_d''} = \frac{231}{769} = 0.3 \text{ }\Omega$$

■

QUESTIONS

4.1 Name the basic parts of a synchronous machine.

4.2 What are the functions of the parts listed in Question 4.1?

4.3 Why is a salient rotor not suitable for a high-speed turboalternator?

4.4 What is the purpose of damper bars in a synchronous machine?

4.5 What are the factors that govern the speed of a synchronous motor? Is the speed dependent on the line voltage? Line current? Line frequency? Size of the motor?

4.6 What is a double-layer winding?

4.7 What is the advantage of a fractional-pitch winding?

4.8 How does fractional pitching affect the induced emf of a synchronous generator?

4.9 What is meant by a distributed winding?

4.10 Under what condition can the synchronous impedance of a synchronous machine be replaced by the synchronous reactance in performance calculations?

4.11 When can the voltage regulation of a synchronous generator be negative?

4.12 What is meant by the power angle of a synchronous machine? In an isolated synchronous machine, does the power angle depend on the power factor angle of the load?

4.13 Will there be a torque developed by an unexcited synchronous motor if the rotor is (a) cylindrical? (b) salient?

4.14 On a given load, is it possible for a synchronous motor to draw the same armature current for two different field currents?

4.15 To operate with a leading power factor, does a synchronous motor have to be overexcited or underexcited?

4.16 What are some of the reasons to operate synchronous machines in parallel?

4.17 Can a cylindrical rotor synchronous machine be operated in parallel with a salient rotor machine?

4.18 What are the conditions for a satisfactory parallel operation of two synchronous generators?

4.19 Does saturation increase or decrease the value of the synchronous reactance of a synchronous machine?

PROBLEMS

4.1 At what speed must a six-pole synchronous generator run to generate 50-Hz voltage?

4.2 A three-phase four-pole 60-Hz synchronous generator has 24 slots. The coil pitch is five slots. Calculate (a) the pitch factor and (b) the distribution factor.

4.3 A three-phase eight-pole 60-Hz synchronous generator has a 60-mWb flux per pole. The winding is full-pitched, and the distribution factor is 0.96. The armature has 120 turns per phase. Calculate the induced voltage per phase.

4.4 The armature of a three-phase eight-pole 900-rpm synchronous generator is wye-connected, and has 72 slots with 10 conductors per slot. The winding factor is 0.96. If 2400 V is measured across the line, determine the flux per pole.

4.5 The armature of a four-pole three-phase 1800 rpm machine is wye-connected and has 48 slots with four conductors per slot wound in two layers. The coil pitch is 150°, and the flux per pole is 80 mWb. Determine the open-circuit line-to-line voltage.

4.6 The open-circuit voltage of a 60-Hz generator is 11,000 V at a field current of 5 A. Calculate the open-circuit voltage at 50 Hz and 2.5 A of field current. Neglect saturation.

4.7 A 60-kVA three-phase wye-connected 440-V 60-Hz synchronous generator has a resistance of 0.15 Ω and a synchronous reactance of 3.5 Ω per phase. At rated load and unity power factor, calculate the percent voltage regulation.

4.8 A 1000-kVA 11-kV three-phase wye-connected synchronous generator supplies a 600-kW 0.8 leading power factor load. The synchronous reactance is 24 Ω per phase and the armature resistance is negligible. Calculate (a) the power angle and (b) the voltage regulation.

4.9 A synchronous generator produces 50 A of short circuit armature current per phase at a field current of 2.3 A. At this field current the open-circuit voltage is 250 V per phase. If the armature resistance is 0.9 Ω per phase, and the generator supplies a purely resistive load of 3 Ω per phase at a 130-V phase voltage and 50 A of armature current, determine the voltage regulation.

4.10 A 1000-kVA 11-kV three-phase wye-connected synchronous generator has an armature resistance of 0.5 Ω and a synchronous reactance of 5 Ω. At a certain field current the generator delivers rated load at 0.9 lagging power factor at 11 kV. For the same excitation, what is the terminal voltage at 0.9 leading power factor full-load?

4.11 An 11-kV three-phase wye-connected generator has a synchronous impedance of 6 Ω per phase, and negligible armature resistance. For a given field current, the open-circuit voltage is 12 kV. Calculate the maximum power developed by the generator. Determine the armature current and power factor for the maximum power condition.

4.12 A 400-V three-phase wye-connected synchronous motor delivers 12 hp at the shaft and operates at 0.866 lagging power factor. The total iron, friction, and field copper losses are 1200 W. If the armature resistance is 0.75 Ω per phase, determine the efficiency of the motor.

4.13 The motor of Problem 4.12 has a synchronous reactance of 6 Ω per phase, and operates at 0.9 leading power factor while taking an armature current of 20 A. Calculate the induced voltage.

4.14 A 1000-kVA 11-kV three-phase wye-connected synchronous motor has a 10-Ω synchronous reactance and a negligible armature resistance. Calculate the induced voltage for (a) 0.8 lagging power factor, (b) unity power factor, and (c) 0.8 leading power factor, when the motor takes 1000 kVA (in each case).

4.15 The per phase induced voltage of a synchronous motor is 2500 V. It lags behind the terminal voltage by 30°. If the terminal voltage is 2200 V per phase, determine the operating power factor. The per phase armature reactance is 6 Ω. Neglect armature resistance.

4.16 The per phase synchronous reactance of a synchronous motor is 8 Ω, and its armature resistance is negligible. The per phase input power is 400 kW and the induced voltage is 5200 V per phase. If the terminal voltage is 3800 V per phase, determine (a) the power factor and (b) the armature current.

4.17 A 2200-V three-phase 60-Hz four-pole wye-connected synchronous motor has a synchronous reactance of 4 Ω and an armature resistance of 0.1 Ω. The excitation is so adjusted that the induced voltage is 2200 V (line to line). If the line current is 220 A at a certain load, calculate (a) the input power, (b) the developed torque, and (c) the power angle.

4.18 An overexcited synchronous motor is connected across a 150-kVA inductive load of 0.7 lagging power factor. The motor takes 12 kW while running on no-load. Calculate the kVA rating of the motor if it is desired to bring the overall power factor of the motor-inductive load combination to unity.

4.19 Repeat Problem 4.18 if the synchronous motor is used to supply a 100-hp load at an efficiency of 90 percent.

4.20 A 60-kVA 400-V three-phase wye-connected salient-pole synchronous generator runs at 75 percent full-load at 0.9 leading power factor. The per phase direct- and quadrature-axis reactances are 1.4 Ω and 0.8 Ω, respectively. Neglecting the armature resistance, calculate (a) the power angle and (b) the developed power.

4.21 Two identical three-phase wye-connected synchronous generators share equally a load of 2500 kW at 33 kV and 0.866 lagging power factor. Each machine has a 5-Ω synchronous reactance per phase and a negligible armature resistance. One of the generators carries 25 A of current at a lagging power factor. What is the current supplied by the second machine?

4.22 The two machines of Problem 4.21 share the load equally. If the power factor of one machine is 0.9 lagging, determine (a) the power factor and (b) the current of the second machine.

CHAPTER 5

Induction Machines

5.1 INTRODUCTION

The induction motor is the most commonly used electric motor. It is considered to be the workhorse of the industry. Like the dc machine and the synchronous machine, an induction machine consists of a stator and a rotor mounted on bearings and separated from the stator by an air gap. Electromagnetically, the stator consists of a core made up of punchings (or laminations) carrying slot-embedded conductors. These conductors are interconnected in a predetermined fashion and constitute the armature windings. These windings are similar to those shown in Fig. 4.6.

Alternating current is supplied to the stator windings, and the currents in the rotor windings are induced by the stator currents. The rotor of the induction machine is cylindrical and carries either (1) conducting bars short-circuited at both ends, as in a *cage-type* machine, or (2) a polyphase winding with terminals brought out to slip rings for external connections, as in a *wound-rotor* machine. A wound-rotor winding is similar to that of the stator. Sometimes the cage-type machine is also called a *brushless* machine and the wound-rotor machine a *slip-ring* machine. The stator and the rotor, in its three different stages of production, are shown in Fig. 5.1. The motor is rated at 2500 kW, 3 kV, 575 A,

FIGURE 5-1. Rotor for a 2500-kW 3 kV 2-pole 400 Hz induction motor in different stages of production. (Courtesy of Brown Boveri Company)

two-pole, and 400 Hz. A finished cage-type rotor of a 3400-kW 6-kV motor is shown in Fig. 5.2, and Fig. 5.3 shows the wound rotor of a three-phase slip-ring 15,200-kW four-pole induction motor. A cutaway view of a completely assembled motor, with a cage-type rotor, is shown in Fig. 5.4. The rotor is housed within the stator and is free to rotate therein.

An induction machine operates on the basis of interaction of induced rotor currents and the air-gap fields. If the rotor is allowed to run under the torque developed by this interaction, the machine will operate as a motor. On the other hand, the rotor may be driven by an external source beyond a speed such that the machine begins to deliver electrical power and operates as an induction generator (instead of as an induction motor, which absorbs electrical power). Thus we see that the induction machine is capable of functioning as a motor as well as a generator. In practice, applications of the induction machine as a generator are less common than motor applications. We will first study the motor operation, then develop the equivalent circuit of an induction motor, and subsequently we will show that the complete characteristics of an induction machine, operating either as a motor or as a generator, are obtainable from the equivalent circuit.

FIGURE 5-2. Complete rotor of a 3400 kW 6 kV 990 rpm induction motor. (Courtesy of Brown Boveri Company)

FIGURE 5-3. Rotor of a 15,200 kW 2.4 kV 3-phase slip-ring induction motor. (Courtesy of Brown Boveri Company)

FIGURE 5-4. Cutaway of a 3-phase cage-type induction motor. (Courtesy of General Electric Company)

5.2

OPERATION OF A THREE-PHASE INDUCTION MOTOR

The key to the operation of an induction motor is the production of the rotating magnetic field. We established in Section 4.3 that a three-phase stator excitation produces a rotating magnetic field in the air gap of the machine, and the field rotates at a synchronous speed given by (4.3). As the magnetic field rotates, it "cuts" the rotor conductors. By this process, voltages are induced in the conductors. The induced voltages give rise to rotor currents, which interact with the air-gap field to produce a torque. The torque is maintained as long as the rotating magnetic field and the induced rotor current exist. Consequently, the rotor starts rotating in the direction of the rotating field. The rotor will achieve a steady-state speed, n, such that $n < n_s$. Clearly, when $n = n_s$, there will be no induced currents and hence no torque. The condition $n > n_s$ corresponds to the generator mode, as we shall see later.

An alternative approach to explaining the operation of the polyphase induction motor is by considering the interaction of the (excited) stator magnetic field with the (induced) rotor magnetic field. The stator excitation produces a rotating magnetic field, which rotates in the air gap at a synchronous speed. The field induces polyphase currents in the rotor, thereby giving rise to another rotating magnetic field, which also rotates at the same synchronous speed as that of the stator and with respect to the stator. Thus we have two rotating magnetic fields, rotating at a synchronous speed with respect to the stator but stationary with respect to each other. Consequently, according to the principle of alignment of magnetic fields, the rotor experiences a torque. The rotor rotates in the direction of the rotating field of the stator.

5.3 SLIP

The actual mechanical speed, n, of the rotor is often expressed as a fraction of the synchronous speed, n_s, as related by *slip*, s, defined as

$$s = \frac{n_s - n}{n_s} \tag{5.1}$$

where n_s is as given in (4.7), which is repeated below for convenience:

$$n_s = \frac{120f}{P} \tag{5.2}$$

The slip may also be expressed as percent slip as follows:

$$\text{percent slip} = \frac{n_s - n}{n_s} \times 100 \tag{5.3}$$

At standstill, the rotating magnetic field produced by the stator has the same relative speed with respect to the rotor windings as with respect to the stator windings. Thus the frequency of the rotor currents, f_r, is the same as the frequency of stator currents, f. At synchronous speed, there is no relative motion between the rotating field and the rotor, and the frequency of rotor current is zero. At other speed, the rotor frequency is proportional to the slip; that is,

$$f_r = sf \tag{5.4}$$

where f_r is the frequency of rotor currents and f is the frequency of stator input current (or voltage).

EXAMPLE 5.1

A six-pole three-phase 60-Hz induction runs at 4 percent slip at a certain load. Determine (a) the synchronous speed, (b) the rotor speed, (c) the frequency of rotor currents, (d) the speed of the rotor rotating field with respect to the stator, and (e) the speed of the rotor rotating field with respect to the stator rotating field.

SOLUTION
(a) From (5.2),

$$n_s = \frac{120 \times 60}{6} = 1200 \text{ rpm}$$

(b) From (5.1),

$$n = (1 - s)n_s = (1 - 0.04) \times 1200 = 1152 \text{ rpm}$$

(c) From (5.4),

$$f_r = 0.04 \times 60 = 2.4 \text{ Hz}$$

(d) The six poles on the stator induce six poles on the rotor. The rotating field produced by the rotor rotates at a corresponding synchronous speed, n_r, relative to the rotor such that

$$n_r = \frac{120 f_r}{P} = \frac{120 f}{P} s = s n_s$$

But the speed of the rotor with respect to the stator is

$$n = (1 - s)n_s$$

Hence the speed of the rotor field with respect to the stator is

$$n'_s = n_r + n = s n_s + (1 - s)n_s = 1200 \text{ rpm}$$

(e) The speed of the rotor field with respect to the stator field is

$$n'_s - n_s = n_s - n_s = 0$$

5.4
DEVELOPMENT OF EQUIVALENT CIRCUITS

In order to develop an equivalent circuit of an induction motor, we consider the similarities between a transformer and an induction motor (on a per phase basis).

If we consider the primary of a transformer to be similar to the stator of the induction motor, its rotor corresponds to the secondary of the transformer. From this analogy, it follows that the stator and rotor each have their respective resistances and leakage reactances. Because the stator and the rotor are magnetically coupled, we must have a magnetizing reactance, just as in a transformer. The air gap in an induction motor makes its magnetic circuit relatively poor and the corresponding magnetizing reactance will be relatively smaller, compared to that of a transformer. The hysteresis and eddy-current losses in an induction motor can be represented by a shunt resistance, as was done for the transformer. Up to this point, we have mentioned the similarities between a transformer and an induction motor. A major difference between the two, however, is introduced because of the rotation of the rotor. Consequently, the frequency of rotor currents is different from the frequency of the stator currents [see (5.4)]. Keeping these facts in mind, we now proceed to represent a three-phase induction motor by a stationary equivalent circuit.

Considering the rotor first and recognizing that the frequency of rotor currents is the slip frequency, we may express the per phase rotor leakage reactance, x_2, at a slip s, in terms of the standstill per phase reactance X_2:

$$x_2 = sX_2 \tag{5.5}$$

Next we observe that the magnitude of the voltage induced in the rotor circuit is also proportional to the slip.

A justification of this statement follows from transformer theory because we may view the induction rotor at standstill as a transformer with an air gap. For the transformer we know that the induced voltage, say E_2, is given by

$$E_2 = 4.44 f N \varphi_m \tag{5.6}$$

But at a slip s, the frequency becomes sf. Substituting this value of frequency into (5.6) yields the voltage e_2 at a slip s as

$$e_2 = 4.44 s f N \varphi_m = sE_2$$

We conclude, therefore, that if E_2 is the per phase voltage induced in the rotor at standstill, the voltage e_2 at a slip s is given by

$$e_2 = sE_2 \tag{5.7}$$

Using (5.5) and (5.7), we obtain the rotor equivalent circuit shown in Fig. 5.5(a). The rotor current I_2 is given by

$$I_2 = \frac{sE_2}{\sqrt{R_2^2 + (sX_2)^2}}$$

which may be rewritten as

$$I_2 = \frac{E_2}{\sqrt{\left(\frac{R_2}{s}\right)^2 + X_2^2}} \qquad (5.8)$$

resulting in the alternative form of the equivalent circuit shown in Fig. 5.5(b). Notice that these circuits are drawn on a per phase basis. To this circuit we may now add the per phase stator equivalent circuit to obtain the complete equivalent circuit of the induction motor.

In an induction motor, only the stator is connected to the ac source. The rotor is not generally connected to an external source, and rotor voltage and current are produced by induction. In this regard, as mentioned earlier, the induction motor may be viewed as a transformer with an air gap, having a variable resistance in the secondary. Thus we may consider that the primary of the transformer corresponds to the stator of the induction motor, whereas the secondary corresponds to the rotor on a per phase basis. Because of the air gap, however, the value of the magnetizing reactance, X_m, tends to be relatively low compared with that of a transformer. As in a transformer, we have a mutual flux linking both the stator and rotor, represented by the magnetizing reactance and various leakage fluxes. For instance, the total rotor leakage flux is denoted by X_2 in Fig. 5.5. Now considering that the rotor is coupled to the stator as the secondary of a transformer is coupled to its primary, we may draw the circuit shown in Fig. 5.6. To develop this circuit further, we need to express the rotor quantities as referred to the stator, as was done for the transformer. The pertinent details are cumbersome and are not considered here. However, having referred the rotor quantities to the stator, we obtain the exact equivalent circuit (per phase) shown in Fig. 5.7 from the circuit given in Fig. 5.6. For reasons that will become immediately clear, we split R_2'/s as

$$\frac{R_2'}{s} = R_2' + \frac{R_2'}{s}(1 - s)$$

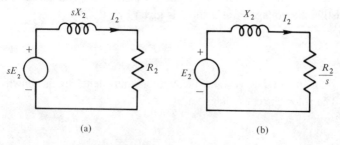

FIGURE 5-5. Two forms of rotor equivalent circuit.

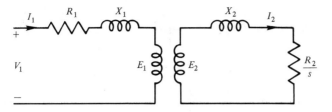

FIGURE 5-6. Stator and rotor as coupled circuits.

to obtain the circuit shown in Fig. 5.7. Here R'_2 is simply the per phase standstill rotor resistance referred to the stator and $R'_2(1 - s)/s$ is a dynamic resistance that depends on the rotor speed and corresponds to the load on the motor. Notice that all the parameters shown in Fig. 5.7 are standstill values and the circuit is the per phase exact equivalent circuit referred to the stator.

5.5

PERFORMANCE CALCULATIONS

We will now show the usefulness of equivalent circuit in determining the motor performance. To illustrate the procedure we refer to Fig. 5.7. We make an approximation in Fig. 5.7, and neglect R_m. We draw the approximate circuit in Fig. 5.8, where we also show approximately the power flow and various power

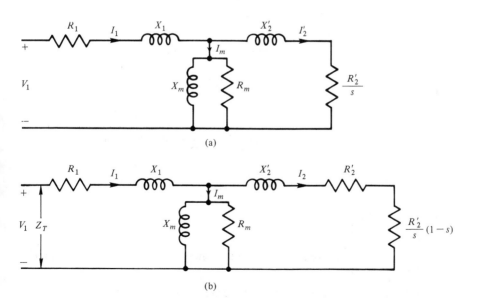

FIGURE 5-7. Two forms of equivalent circuits of an induction motor.

5.5 PERFORMANCE CALCULATIONS

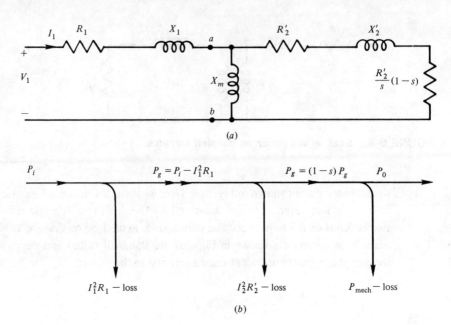

FIGURE 5-8. (a) An approximate equivalent circuit of an induction motor; (b) power flow in an induction motor.

losses in one phase of the machine. From Fig. 5.8 we obtain the following relationships on a per phase basis:

$$\text{stator } I^2R \text{ loss} = I_1^2 R_1 \tag{5.9a}$$

$$\text{power crossing the air gap} = P_g = P_i - I_1^2 R_1 \tag{5.9b}$$

Since this power, P_g, is dissipated in R_2'/s [of Fig. 5.7(a)], we also have

$$P_g = \frac{I_2^2 R_2'}{s} \tag{5.10}$$

Subtracting $I_2^2 R_2'$ loss from P_g yields the developed electromagnetic power, P_d. Thus

$$P_d = P_g - I_2^2 R_2' \tag{5.11}$$

From (5.10) and (5.11) we get

$$P_d = (1 - s)P_g \tag{5.12}$$

This power appears across the resistance $R_2'(1 - s)/s$ (of Fig. 5.8) corresponding to the load. Subtracting the mechanical rotational power, P_r, from P_d gives the output power, P_o. Hence

$$P_o = P_d - P_r \tag{5.13}$$

and

$$\text{efficiency} = \frac{P_o}{P_i} \tag{5.14}$$

Torque calculations can be made from the power calculations. Thus, to determine the electromagnetic torque, T_e, developed by the motor, at a speed ω_m (rad/s), we write

$$T_e \omega_m = P_d \tag{5.15}$$

But

$$\omega_m = (1 - s)\omega_s \tag{5.16}$$

where ω_s is the synchronous speed in rad/s. From (5.12), (5.15), and (5.16) we obtain

$$T_e = \frac{P_g}{\omega_s} \tag{5.17}$$

which gives the torque at a slip s. At standstill $s = 1$; hence the standstill torque developed by the motor is given by

$$(T_e)_{\text{standstill}} = \frac{P_{gs}}{\omega_s} \tag{5.18}$$

Notice that in Fig. 5.8 we have neglected the core losses, most of which are in the stator. We will include core losses only in efficiency calculations. The reason for this simplification is to reduce the amount of the complex arithmetic required in numerical computations. We now illustrate the details of calculations by the following example.

EXAMPLE 5.2
The parameters of the equivalent circuit (Fig. 5.8) for a 220-V three-phase four-pole wye-connected 60-Hz induction motor are

$$R_1 = 0.2 \, \Omega \qquad R_2' = 0.1 \, \Omega$$

$$X_1 = 0.5 \, \Omega \qquad X_2' = 0.2 \, \Omega$$

$$X_m = 20.0 \, \Omega$$

The total iron and mechanical losses are 350 W. For a slip of 2.5 percent, calculate input current, output power, output torque, and efficiency.

SOLUTION

From Fig. 5.8, the total impedance is

$$Z_1 = R_1 + jX_1 + \frac{jX_m\left(\frac{R_2'}{s} + jX_2'\right)}{\frac{R_2'}{s} + j(X_m + X_2')}$$

$$= 0.2 + j0.5 + \frac{j20(4 + j0.2)}{4 + j(20 + 0.2)}$$

$$= (0.2 + j0.5) + (3.77 + j0.95) = 4.23 \underline{/20°}$$

$$V_1 = \text{phase voltage} = \frac{220}{\sqrt{3}} = 127 \text{ V}$$

$$I_1 = \text{input current} = \frac{127}{4.23} = 30 \text{ A}$$

$$\cos \varphi = \text{power factor} = \cos 20° = 0.94$$

$$3V_1I_1 \cos \varphi = \text{total input power} = \sqrt{3} \times 220 \times 30 \times 0.94 = 10.75 \text{ kW}$$

From the equivalence of Figs. 5.8(a) and 5.9, we obtain total power across the air gap

$$P_g = 3 \times 30^2 \times 3.77 = 10.18 \text{ kW}$$

Notice that 3.77 Ω is the equivalent resistance between terminals a and b in the two circuits.

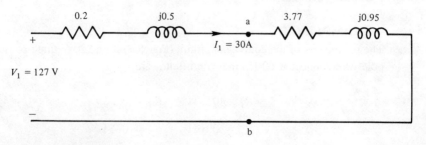

FIGURE 5-9. Example 5.2.

total power developed, $P_d = (1-s)P_g = 0.975 \times 10.18 = 9.93$ kW

total output power $= P_d - P_{core} = 9.93 - 0.35 = 9.58$ kW

$$\text{total output torque} = \frac{\text{output power}}{\omega_m}$$

$$= \frac{9.58}{184} \times 1000 = 52 \text{ N-m}$$

where, from (4.6)

$$\omega_m = (1-s)\frac{\omega}{p} = 0.075 \times 60 \times \pi = 184 \text{ rad/s}$$

$$\text{efficiency} = \frac{\text{output power}}{\text{input power}} = \frac{9.58}{10.75} = 89.1\%$$ ∎

Using this procedure, we can calculate the performance of the motor at other values of the slip, ranging from 0 to 1. The characteristics thus calculated are shown in Fig. 5.10.

EXAMPLE 5.3
A two-pole 60-Hz three-phase induction motor develops 25 kW of electromagnetic power at a certain speed. The rotational mechanical loss at this speed is 400 W. If the power crossing the air gap is 27 kW, calculate (a) the slip and (b) the output torque.

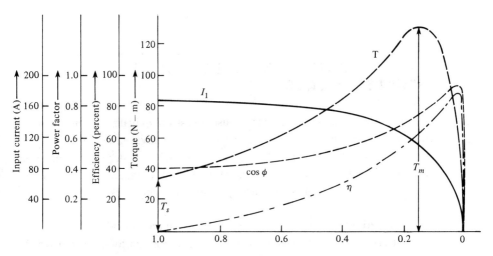

FIGURE 5-10. Characteristics of an induction motor. T_m = maximum torque; T_s = starting torque.

5.5 PERFORMANCE CALCULATIONS

SOLUTION
(a) From (5.11) we have

$$1 - s = \frac{P_d}{P_g}$$

or

$$1 - s = \frac{25}{27}$$

Thus $s = 0.074$ (or 7.4 percent).

(b) The developed torque is given by (5.17). Substituting $\omega_s = 2\pi n_s/60 = 2\pi \times 3600/60$, $P_g = 27$ kW (given) in (5.17) gives

$$T_e = \frac{27,000}{2\pi \times 3600/60} = 71.62 \text{ N-m}$$

Torque lost due to mechanical rotation is found from

$$T_{\text{loss}} = \frac{P_{\text{loss}}}{\omega_m} = \frac{400}{(1 - 0.074)2\pi \times 3600/60} = 1.15 \text{ N-m}$$

Hence

$$\text{output torque} = T_e - T_{\text{loss}} = 71.62 - 1.15 = 70.47 \text{ N-m} \quad \blacksquare$$

5.6
APPROXIMATE EQUIVALENT CIRCUIT FROM TEST DATA

The preceding examples illustrate the usefulness of equivalent circuits of induction motors. For most purposes, an approximate equivalent circuit is adequate. One such circuit is shown is Fig. 5.12. Obviously, in order to use this circuit for calculations, its parameters must be known. The parameters of the circuit shown in Fig. 5.12 can be obtained from the following tests.

No-Load Test. In this test, the motor is run on no-load. Input power and current are recorded at several voltages ranging from 25 to 125 percent of the rated voltage. Results thus obtained are plotted in Fig. 5.11. As the voltage is reduced, the flux decreases proportionately. The power curve is approximately

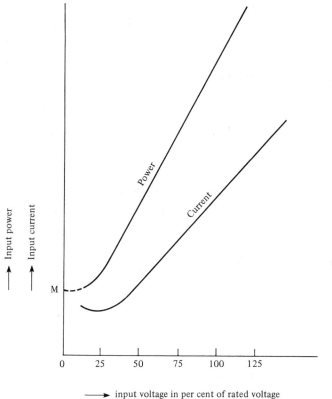

FIGURE 5-11. No-load test data.

parabolic. At about 20 percent of the rated voltage, the magnetizing current and core loss are small, and the input power is almost entirely due to mechanical losses. By extrapolating the power curve to zero voltage, as shown in Fig. 5.11, we obtain the intercept OM which corresponds to mechanical (or friction and windage) loss. At the rated voltage, let the per phase input power (after correcting for the friction and windage loss, as mentioned above), voltage, and current be denoted by P_0, V_0, and I_0, respectively. When the machine runs on no-load, the slip is close to zero and the circuit in Fig. 5.12 to the right of the shunt branch is taken to be an open circuit. Thus the parameters R_m and X_m are found from

$$R_m = \frac{V_0^2}{P_0} \qquad (5.19)$$

$$X_m = \frac{V_0^2}{\sqrt{V_0^2 I_0^2 - P_0^2}} \qquad (5.20)$$

Blocked-Rotor Test. In this test, the rotor of the machine is blocked ($s = 1$), and a reduced voltage is applied to the machine so that the rated current flows through the stator windings. The input power, voltage, and current are recorded and reduced to per phase values; these are denoted by P_s, V_s, and I_s, respectively. In this test, the iron losses are assumed to be negligible and the shunt branch of the circuit shown in Fig. 5.12 is considered to be absent. The parameters are thus found from

$$R_e = R_1 + a^2 R_2 = \frac{P_s}{I_s^2} \quad (5.21)$$

$$X_e = X_1 + a^2 X_2 = \frac{\sqrt{V_s^2 I_s^2 - P_s^2}}{I_s^2} \quad (5.22)$$

In (5.21) and (5.22), the constant a is the turns ratio. The stator resistance per phase, R_1, can be directly measured, and knowing R_e from (5.21), we can determine $R_2' = a^2 R_2$, the rotor resistance referred to the stator. There is no simple method of determining X_1 and $X_2' = a^2 X_2$ separately. The total value given by (5.22) is sometimes equally divided between X_1 and X_2'.

EXAMPLE 5.4

The results of no-load and blocked-rotor tests on a three-phase wye-connected induction motor are as follows:

No-load test:

line-to-line voltage = 400 V

input power = 1770 W

input current = 18.5 A

friction and windage loss = 600 W

Blocked-rotor test:

line-to-line voltage = 45 V

input power = 2700 W

input current = 63 A

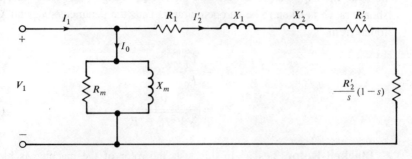

FIGURE 5-12. An approximate equivalent circuit (per phase) of an induction motor.

Determine the parameters of the approximate equivalent circuit (Fig. 5.12).

SOLUTION
From no-load test data:

$$V_0 = \frac{400}{\sqrt{3}} = 231 \text{ V} \qquad P_0 = \tfrac{1}{3}(1770 - 600) = 390 \text{ W} \qquad I_0 = 18.5 \text{ A}$$

Then, by (5.19) and (5.20),

$$R_m = \frac{(231)^2}{390} = 136.8 \text{ }\Omega$$

$$X_m = \frac{(231)^2}{\sqrt{(231)^2(18.5)^2 - (390)^2}} = 12.5 \text{ }\Omega$$

From blocked-rotor test data:

$$V_s = \frac{45}{\sqrt{3}} = 25.98 \text{ V} \qquad I_s = 63 \text{ A} \qquad P_s = \frac{2700}{3} = 900 \text{ W}$$

Then, by (5.21) and (5.22),

$$R_e = R_1 + a^2 R_2 = \frac{900}{(63)^2} = 0.23 \text{ }\Omega$$

$$X_e = X_1 + a^2 X_2 = \frac{\sqrt{(25.98)^2(63)^2 - (900)^2}}{(63)^2} = 0.34 \text{ }\Omega \qquad \blacksquare$$

5.7
PERFORMANCE CRITERIA OF INDUCTION MOTORS

Example 5.2 shows the usefulness of the equivalent circuit in calculating the performance of the motor. The performance of an induction motor may be characterized by the following major factors.

1. Efficiency.
2. Power factor.
3. Starting torque.
4. Starting current.
5. Pull-out (or maximum) torque.

These characteristics are shown in Fig. 5.10. In design considerations, heating because of I^2R losses and core losses and means of heat dissipation must be included. It is not within the scope of this book to present a detailed discussion of the effects of design changes and, consequently, parameter variations, on each performance characteristics. Here we summarize the results as trends. For example, the efficiency of an induction motor is proportional to $(1 - s)$. Thus the motor would be most compatible with a load running at the lowest slip. Because the efficiency is clearly dependent on I^2R losses, R_2' and R_1 must be small for a given load. To reduce core losses, the working flux density (B) must be small. But this imposes a conflicting requirement on the load current (I_2') because the torque is dependent on the product of B and I_2'. In other words, an attempt to decrease the core losses beyond a limit would result in an increase in the I^2R losses for a given load.

It may be seen from the equivalent circuits (developed in Section 5.4) that the power factor can be improved by decreasing the leakage reactances and increasing the magnetizing reactance. However, it is not wise to reduce the leakage reactances to a minimum, since the starting current of the motor is essentially limited by these reactances. Again, we notice the conflicting conditions for a high power factor and a low starting current. Also, the pull-out torque would be higher for lower leakage reactances.

A high starting torque is produced by a high R_2'; that is, the higher the rotor resistance, the higher would be the starting torque. A high R_2' is in conflict with a high-efficiency requirement. The effect of varying rotor resistance on the motor torque–speed characteristics is shown in Fig. 5.13, which also shows three

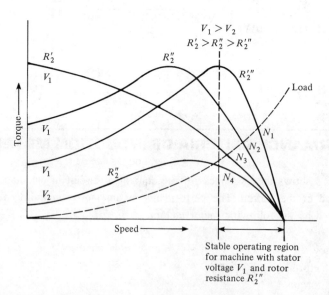

FIGURE 5-13. Effect of rotor resistance on torque-speed characteristics.

different steady-state operating speeds for three values of the rotor resistance and two stator voltages V_1 and V_2.

5.8 SPEED CONTROL OF INDUCTION MOTORS

Because of its simplicity and ruggedness, the induction motor finds numerous applications. However, it suffers from the drawback that, in contrast to dc motors, its speed cannot be easily and efficiently varied continuously over a wide range of operating conditions. We will briefly review the various possible methods by which the speed of the induction motor can be varied either continuously or in discrete steps. We will not consider all these methods in detail here. Certain details are given in Chapter 7.

The speed of the induction motor can be varied by (1) varying the synchronous speed of the rotating field, or (2) varying the slip. Because the efficiency of the induction motor is approximately proportional to $(1 - s)$, any method of speed control that depends on the variation of slip is inherently inefficient. On the other hand, if the supply frequency is constant, varying the speed by changing the synchronous speed results only in discrete changes in the speed of the motor. We will now consider the governing principles of these methods of speed control.

Recall from (5.2) that the synchronous speed n_s of the rotating field in an induction machine is given by

$$n_s = 120fP$$

where P is the number of poles and f is the supply frequency, which indicates that n_s can be varied by (1) changing the number of poles P, or (2) changing the frequency f. Both methods have found applications, and we consider here the pertinent qualitative details.

Pole-Changing Method. In this method the stator winding of the motor is so designed that by changing the connections of the various coils (the terminals of which are brought out), the number of poles of the winding can be changed in the ratio of 2:1. Accordingly, two synchronous speeds result. We observe that only two speeds of operation are possible. Figure 5.14 shows one phase interconnection for a 2:1 pole ratio. If more independent windings (e.g., two) are provided—each arranged for pole changing—more synchronous speeds (e.g., four) can be obtained. However, the fact remains that only discrete changes in the speed of the motor can be obtained by this technique. The method has the

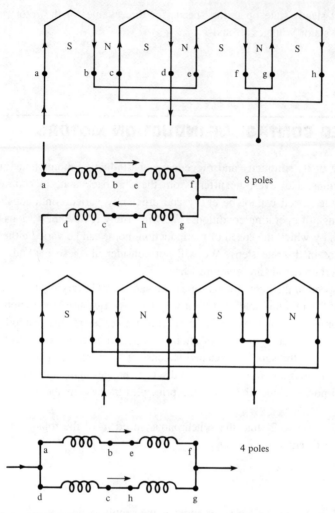

FIGURE 5-14. Pole-changing (2/1)

advantage of being efficient and reliable because the motor has a squirrel-cage rotor and no brushes.

Variable-Frequency Method. We know that the synchronous speed is directly proportional to the frequency. If it is practicable to vary the supply frequency, the synchronous speed of the motor can also be varied. The variation in speed is continuous or discrete according to continuous or discrete variation of the supply frequency. However, the maximum torque developed by the motor is inversely proportional to the synchronous speed. If we desire a constant maximum torque, both supply voltage and supply frequency should be increased if we wish to increase the synchronous speed of the motor. The inherent difficulty in the application of this method is that the supply frequency, which is commonly available, is fixed. Thus the method is applicable only if a variable-frequency supply is available. Various schemes have been proposed to obtain a variable-frequency supply. With the advent of solid-state devices with comparatively large power ratings, it is now possible to use static inverters to drive the induction motor.

Variable-Slip Method. Controlling the speed of an induction motor by changing its slip may be understood by reference to Fig. 5.13. The dashed curve shows the speed–torque characteristic of the load. The curves with solid lines are the speed–torque characteristics of the induction motor under various conditions (such as different rotor resistances—R'_2, R''_2, R'''_3— or different stator voltages—V_1, V_2). We have four different torque–speed curves and, therefore, the motor can run at any one of four speeds—N_1, N_2, N_3, and N_4—for the given load. Note that to the right of the peak torque is the stable operating region of the motor. In practice, the slip of the motor can be changed by one of the following methods.

Variable Stator Voltage Methods. Since the electromagnetic torque developed by the machine is proportional to the square of the applied voltage, we obtain different torque–speed curves for different voltages applied to the motor. For a given rotor resistance, R_2, two such curves are shown in Fig. 5.13 for two applied voltages V_1 and V_2. Thus the motor can run at speeds N_2 or N_4. If the voltage can be varied continuously from V_1 to V_2, the speed of the motor can also be varied continuously between N_2 and N_4 for the given load. This method is applicable to cage-type as well as wound-rotor-type induction motors.

Variable Rotor-Resistance Method. This method is applicable only to the wound-rotor motor. The effect on the speed–torque curves of inserting external resistances in the rotor circuit is shown in Fig. 5.13 for three different rotor resistances R'_2, R''_2, and R'''_2. For the given load, three speeds of operation are possible. Of course, by continuous variation of the rotor resistance, continuous variation of the speed is possible.

Control by Solid-State Switching. Other than the inverter-driven motor, the speed of the wound-rotor motor can be controlled by inserting the inverter in the rotor circuit or by controlling the stator voltage by means of solid-state switching devices such as silicon-controlled rectifiers (SCRs or thyristors). The output from the SCR feeding the motor is controlled by adjusting its firing angle. The method of doing this is similar to the variable-voltage method outlined earlier. However, it has been found that control by an SCR gives a wider range of operation and is more efficient than other slip-control methods. For details, see Chapter 7.

5.9 STARTING OF INDUCTION MOTORS

Most induction motors—large and small—are rugged enough that they could be started across the line without incurring any damage to the motor windings, although about five to seven times the rated current flows through the stator at rated voltage at standstill. However, in large induction motors, large starting currents are objectionable in two respects. First, the mains supplying the induction motor may not be of a sufficiently large capacity. Second, because of a large starting current, the voltage drops in the lines may be excessive resulting in a reduced voltage across the motor. Because the torque varies approximately as the square of the voltage, the starting torque may become so small, at the reduced line voltage, that the motor might not even start on load. Thus, we formulate the basic requirement for starting: the line current should be limited by the capacity of the mains, but only to the extent that the motor can develop sufficient torque to start (on load, if necessary).

EXAMPLE 5.5
An induction motor is designed to run at 5 percent slip on full-load. If the motor draws six times the full-load current at starting at the rated voltage, estimate the ratio of starting torque to the full-load torque.

SOLUTION
The torque at a slip s is given by (5.17), which in conjunction with (5.10) becomes

$$T_e = \frac{I_2^2 R_2'}{s\omega_s}$$

At full-load, with $I_2 = I_{2f}$, the torque is

$$T_{ef} = \frac{I_{2f}^2 R_2'}{0.05\omega_s}$$

At starting, $I_{2s} = 6I_{2f}$ and $s = 1$, so that

$$T_{es} = \frac{(6I_{2f})^2 R_2'}{\omega_s}$$

Hence

$$\frac{T_{es}}{T_{ef}} = \frac{(6I_{2f})^2 R_2'}{\omega_s} \frac{0.05\omega_s}{I_{2f}^2 R_2'} = 1.8 \qquad \blacksquare$$

EXAMPLE 5.6

If the motor of the Example 5.5 is started at a reduced voltage to limit the line current to three times the full-load current, what is the ratio of the starting torque to the full-load torque?

SOLUTION
In this case we have

$$\frac{T_{es}}{T_{ef}} = 3^2 \times 0.05 = 0.45$$

■

Notice that the starting torque has reduced by a factor of 4 compared to the case of full-voltage starting. In many practical cases, the line current is limited to six times the full-load current and the starting torque is desired to be about 1.5 times the full-load torque.

There are numerous types of pushbutton starters for induction motors now commercially available. In the following, however, we will briefly consider only the principles of the two commonly used methods. We consider the current-limiting types first. Some of the common methods of limiting the stator current while starting are:

1. *Reduced-voltage starting.* At the time of starting, a reduced voltage is applied to the stator and the voltage is increased to the rated value when the motor is within 25 percent of its final speed. This method has the obvious limitation that a variable-voltage source is needed and the starting torque drops substantially. The wye–delta method of starting is a reduced-voltage starting method. If the stator is normally connected in delta, reconnection to wye reduces the phase voltage, resulting in less current at starting. For example, at starting, if the line current is about five times the full-load current in a delta-connected stator, the current in the wye connection will be less than twice the full-load value. But at the same time, the starting torque for a wye connection would be about one-third its value for a delta connection. The advantage of wye–delta starting is that it is inexpensive and requires only a three-pole (or three single-pole) double-throw switch (or switches), as shown in Fig. 5.15.

2. *Current limiting by series resistance.* Series resistances inserted in the three lines sometimes are used to limit the starting current. These resistances are shorted out when the motor has gained speed. This method has the obvious disadvantage of being inefficient because of the extra losses in the external resistances during the starting period.

Turning now to the starting torque, we recall from the preceding section that the starting torque is dependent on the rotor resistance. Thus a high rotor resistance results in a high starting torque. Therefore, in a wound-rotor machine (see Fig. 5.16), external resistance in the rotor circuit may be conveniently used. In a cage rotor, deep slots are used, where the slot depth is two or three times

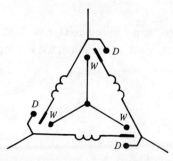

FIGURE 5-15. Wye-delta starting. Switches on **W** correspond to wye and switches on **D** correspond to the delta connection.

greater than the slot width (see Fig. 5.17). Rotor bars embedded in deep slots provide a high effective resistance and a large torque at starting. Under normal running conditions with low slips, however, the rotor resistance becomes lower and efficiency high. This characteristic of rotor bar resistance is a consequence of *skin effect*. Because of skin effect, the current will have a tendency to concentrate at the top of the bars at starting, when the frequency of rotor currents is high. At this point, the frequency of rotor currents will be the same as the stator input frequency (e.g., 60 Hz). While running, the frequency of rotor currents (= slip frequency = 3 Hz at 5 percent slip and 60 Hz) is much lower. At this level of operation, skin effect is negligible and the current almost uniformly distributes throughout the entire bar cross section.

Skin effect is used in an alternative form in a *double-cage* rotor (Fig. 5.18), where the inner cage is deeply embedded in iron and has low-resistance bars and has a high reactance. The outer cage has relatively high resistance bars close to the stator, and has a low reactance of a normal single-cage rotor. At starting, because of skin effect, the influence of the outer cage dominates, thus producing

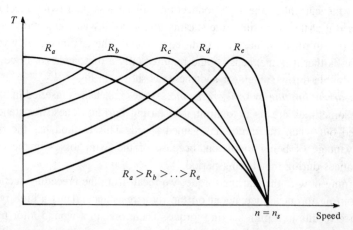

FIGURE 5-16. Effect of changing rotor resistance on the starting of a wound-rotor motor.

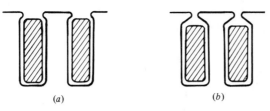

FIGURE 5-17. Deep-bar rotor slots: (a) open; (b) partially closed.

a high starting torque. While running, the current penetrates to full depth into the lower cage—because of insignificant skin effect and lower reactance—which results in an efficient steady-state operation. Notice that under normal running conditions both cages carry current, thus somewhat increasing the power rating of the motor. The rotor equivalent circuit of a double-cage rotor then becomes as shown in Fig. 5.19. Approximately, the cages may be considered to develop separate torques, and the sum of these torques is the total torque. By appropriate designs, the inner- and outer-cage resistances and leakage reactances may be modified to obtain a wide range of performance characteristics. Compared to a normal single-cage motor, the inner-cage leakage reactance in a double-cage motor lowers its power factor at full-load. Furthermore, the high resistance of the outer cage increases the full-load I^2R loss and decreases the motor efficiency. But for a duty cycle of frequent starts and stops, a double-cage motor is likely to have a higher energy efficiency compared to a single-cage motor.

EXAMPLE 5.7

A motor employs a wye–delta starter which connects the motor phases in wye at the time of starting and in delta when the motor is running. The full-load slip is 4 percent and the motor draws nine times the full-load current if started directly from the mains. Determine the ratio of starting torque to full-load torque.

SOLUTION

When the phases are switched to delta, the phase voltage, and hence the full-load current, is increased by a factor of $\sqrt{3}$ over the value it would have had in a wye connection. Then it follows from the last equation in Example 5.6 that

$$\frac{T_s}{T_{FL}} = \left(\frac{9}{\sqrt{3}}\right)^2 (0.04) = 1.08$$

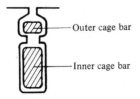

FIGURE 5-18. Form of a slot for a double-cage rotor.

5.9 STARTING

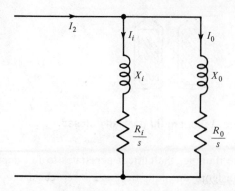

FIGURE 5-19. Equivalent circuit of a double-cage rotor.

EXAMPLE 5.8

To obtain a high starting torque in a cage-type motor, a double-cage rotor is used. The forms of a slot and of the bars of the two cages are shown in Fig. 5.18. The outer cage has a higher resistance than the inner cage. At starting, because of the skin effect, the influence of the outer cage dominates, thus producing a high starting torque. An approximate equivalent circuit for such a rotor is given in Fig. 5.19. Suppose that, for a certain motor, we have the per phase values, at standstill,

$$R_i = 0.1\ \Omega \qquad R_o = 1.2\ \Omega \qquad X_i = 2\ \Omega \qquad X_o = 1\ \Omega$$

Determine the ratio of the torques provided by the two cages at (a) starting and (b) 2 percent slip.

SOLUTION
(a) From Fig. 5.17, at $s = 1$:

$$Z_i^2 = (0.1)^2 + (2)^2 = 4.01\ \Omega^2$$

$$Z_o^2 = (1.2)^2 + (1)^2 = 2.44\ \Omega^2$$

power input to the inner cage $\equiv P_{ii} = I_i^2 R_i = 0.1 I_i^2$

power input to the outer cage $\equiv P_{io} = I_o^2 R_o = 1.2 I_o^2$

$$\frac{\text{torque due to inner cage}}{\text{torque due to outer cage}} = \frac{T_i}{T_o} = \frac{P_{ii}}{P_{io}} = \frac{0.1}{1.2}\left(\frac{I_i}{I_o}\right)^2 = \frac{0.1}{1.2}\left(\frac{Z_o}{Z_i}\right)^2$$

$$= \frac{0.1}{1.2}\left(\frac{2.44}{4.01}\right) = 0.05$$

(b) Similarly, at $s = 0.02$:

$$Z_i^2 = \left(\frac{0.1}{0.02}\right)^2 + (2)^2 = 29 \;\Omega^2$$

$$Z_o^2 = \left(\frac{1.2}{0.02}\right)^2 + (1)^2 = 3601 \;\Omega^2$$

$$\frac{T_i}{T_o} = \frac{0.1}{1.2}\left(\frac{3601}{29}\right) = 10.34$$

∎

5.10 INDUCTION GENERATORS

Up to this point, we have studied the behavior of an induction machine operating as a motor. We recall from the preceding discussions that for motor operation the slip lies between zero and unity, and for this case we have a conversion of electrical power into mechanical power. If the rotor of an induction machine is driven by an auxiliary means such that the rotor speed, n, becomes greater than the synchronous speed, n_s, we have, from (5.1), a negative slip. A negative slip implies that the induction machine is now operating as an induction generator. Alternatively, we may refer to the rotor portion of the equivalent circuit, such as that of Fig. 5.12. If the slip is negative, the resistance representing the load becomes $R_2'[1 - (-s)]/(-s)$, which results in a negative value of the resistance. Because a positive resistance absorbs electrical power, a negative resistance may be considered as a source of power. Hence a negative slip corresponds to a generator operation.

To understand the generator operation, we consider a three-phase induction machine to which a prime mover is coupled mechanically. When the stator is excited, a synchronously rotating magnetic field is produced and the rotor begins to run, as in an induction motor, while drawing electrical power from the supply. The prime mover is then turned on (to rotate the rotor in the direction of the rotating field). When the rotor speed exceeds synchronous speed, the direction of electrical power reverses. The power begins to flow into the supply as the machine begins to operate as a generator. The rotating magnetic field is produced by the magnetizing current supplied to the stator winding from the three-phase source. This supply of the magnetizing current must be available as the machine operates as an induction generator. For induction generators operating in parallel with a three-phase source capable of supplying the necessary exciting current, the voltage and the frequency are fixed by the operating voltage and frequency of the source supplying the exciting current.

An induction generator may be self-excited by providing the magnetizing reactive power by a capacitor bank. In such a case an external ac source is not needed. The generator operating frequency and voltage are determined by the speed of the generator, its load, and the capacitor rating. As for the dc shunt generator, for the induction generator to self-excite, its rotor must have sufficient remanent flux. The operation of a self-excited induction generator may be understood by referring to Fig. 5.20(a). On no-load, the charging current of the capacitor, $I_c = V_1/X_c$, must be equal to the magnetizing current, $I_m = V_1/X_m$. Because V_1 is a function of I_m, for a stable operation the line $I_c X_c = I_m X_c$ must intersect the magnetization curve, which is a plot of V_1 versus I_m, as shown in

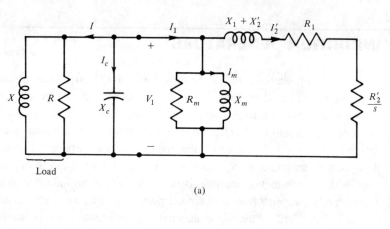

(a)

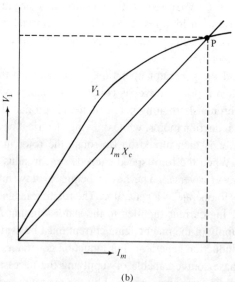

(b)

FIGURE 5-20. (a) Equivalent circuit of a self-excited induction generator; (b) determination of stable operating point.

Fig. 5.20(b). The operating point P is thus determined, and we have

$$V_1 = I_m X_c \qquad (5.23)$$

Since $X_c = 1/\omega C = 1/2\pi fC$, we rewrite (5.23) as

$$I_m = 2\pi fCV_1 \qquad (5.24)$$

From (5.24) the operating frequency is given by

$$f = \frac{I_m}{2\pi CV_1} \qquad (5.25)$$

On load, the generated power $V_1 I_2' \cos \varphi_2'$ provides for the power loss in R_m and the power utilized by the load R. The reactive currents are related to each other by

$$\frac{V_1}{X_c} = \frac{V_1}{X} + \frac{V_1}{X_m} + I_2' \sin \varphi_2' \qquad (5.26)$$

which determines the capacitance for a given load.

Operating characteristics of an induction machine, for generator and motor modes, are shown in Fig. 5.21. Unlike in a synchronous generator, for a given load, the output current and the power factor are determined by the generator parameters. Therefore, when an induction generator delivers a certain power, it also supplies a certain in-phase current and a certain quadrature current. How-

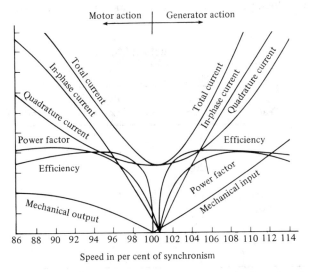

FIGURE 5-21. Motor and generator characteristics of an induction machine.

5.10 INDUCTION GENERATORS

ever, the quadrature component of the current generally does not have a definite relationship to the quadrature component of the load current. The quadrature current must be supplied by the synchronous generators operating in parallel with the induction generator.

Induction generators are not suitable for supplying loads having low lagging power factors. In the past, induction generators have been used in variable-speed constant-frequency generating systems. Large induction generators have found applications in hydroelectric power stations. Induction generators are promising for windmill applications and are rapidly finding applications in this area.

5.11

ENERGY-EFFICIENT INDUCTION MOTORS

Over the last 10 years the cost of electrical energy has more than doubled. For instance, it has been reported that the annual energy cost to operate a 10-hp induction motor 4000 h per year has increased from $850 in 1972 to $1950 in 1980. The escalation of oil prices in the mid-1970s led the manufacturers of electric motors to seek methods to improve motor efficiencies. In order to improve the motor efficiency, its loss distribution must be studied. For a typical standard (NEMA design B; see Chapter 8 for explanation) three-phase 50-hp motor, the loss distribution at full-load is given in Table 5.1. In this table we also show the average loss distribution in percent of total losses for standard induction motors. The per unit loss in Table 5.1 is defined as loss/(hp × 746).

In improving the efficiency of the motor, we must design to achieve a balance among the various losses and, at the same time, meet other specifications, such

TABLE 5.1 Loss Distribution in Standard Induction Motors

Loss Distribution	50-hp Motor			Average Percent Loss for Standard Motors
	Watts	Percent Loss	Per Unit Loss	
Stator I^2R loss	1,540	38	0.04	37
Rotor I^2R loss	860	22	0.02	18
Magnetic core loss	765	20	0.02	20
Friction and windage loss	300	8	0.01	9
Stray load loss	452	12	0.01	16
Total losses	3,917	100	0.10	
Output (W)	37,300			
Input (W)	41,217			
Efficiency (%)	90.5			

as breakdown torque, locked-rotor current and torque, and power factor. For the motor designer, a clear understanding of the loss distribution is very important. Loss reductions can be made by increasing the amount of the material in the motor. Without making other major design changes, a loss reduction of about 10 percent at full load can be achieved. Improving the magnetic circuit design using lower-loss electrical grade laminations can result in a further reduction of losses by about 10 percent. The cost of improving the motor efficiency increases with output rating (hp) of the motor. Based on the improvements just mentioned to increase the motor efficiency, Fig. 5.22 shows a comparison between the efficiencies of energy-efficient motors and those of standard motors.

Several of the major manufacturers of induction motors have developed product lines of energy-efficient motors. These motors are identified by their trade names, such as:

E-Plus (Gould Inc.).
Energy Saver (General Electric).
XE–Energy Efficient (Reliance Electric).
Mac II High Efficiency (Westinghouse).

Because energy-efficient motors use more material, they are relatively bigger in size compared to standard motors.

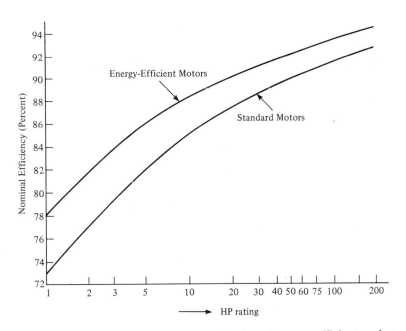

FIGURE 5-22. A comparison of nominal efficiencies of energy efficient and standard motors.

5.11 ENERGY-EFFICIENT MOTORS

QUESTIONS

5.1 What are the two types of rotors used in an induction motor? What rotor type is used in a brushless induction motor?

5.2 Why is it necessary to have a rotating magnetic field for the operation of an induction motor?

5.3 What is meant by slip?

5.4 How are rotor currents produced in an induction motor?

5.5 How is the frequency of rotor currents related to the frequency of stator currents?

5.6 What is the effect of rotor resistance on the developed torque of an induction motor?

5.7 How does the supply frequency affect the speed of an induction motor?

5.8 In a dc motor and in a synchronous motor, stator and rotor windings are both connected to power sources. For the operation of an induction motor, is it necessary to connect both stator and rotor to an ac source?

5.9 Name the parts of an induction motor where most of the losses occur.

5.10 Name the tests commonly performed to obtain the parameters of the approximate equivalent circuit of an induction motor.

5.11 What measures may be taken to obtain a high starting torque from an induction motor?

5.12 How can the speed of an induction motor be controlled?

5.13 What factors influence the starting of an induction motor?

5.14 What is the primary function of a double-cage rotor?

5.15 How can an induction machine be operated as an induction generator?

5.16 Does the frequency of the voltage of an induction generator depend on its speed?

5.17 Is an induction generator self-exciting (like a dc shunt generator)?

PROBLEMS

5.1 A six-pole 60-Hz induction motor runs at 1152 rpm. Determine the synchronous speed and the percent slip.

5.2 A six-pole induction motor is supplied by a synchronous generator having four poles and running at 1500 rpm. If the speed of the induction motor is 750 rpm, what is the frequency of rotor current?

5.3 A three-phase four-pole 60-Hz induction motor develops a maximum torque of 180 N-m at 800 rpm. If the rotor resistance is 0.2 Ω per phase, determine the developed torque at 1000 rpm.

5.4 A 400-V four-pole three-phase 60-Hz induction motor has a wye-connected rotor having an impedance of $(0.1 + j0.5)$ Ω per phase. How much additional resistance must be inserted in the rotor circuit for the motor to develop the maximum starting torque? The effective stator-to-rotor turns ratio is 1.

5.5 For the motor of Problem 5.4, what is the motor speed corresponding to the maximum developed torque without any external resistance in the rotor circuit?

5.6 A two-pole 60-Hz induction motor develops a maximum torque of twice the full-load torque. The starting torque is equal to the full-load torque. Determine the full-load speed.

5.7 The input to the rotor circuit of a four-pole 60-Hz induction motor, running at 1000 rpm, is 3 kW. What is the rotor copper loss?

5.8 The stator current of a 400-V three-phase wye-connected four-pole 60-Hz induction motor running at a 6 percent slip is 60 A at 0.866 power factor. The stator copper loss is 2700 W and the total iron and rotational losses are 3600 W. Calculate the motor efficiency.

5.9 An induction motor has an output of 30 kW at 86 percent efficiency. For this operating condition, stator copper loss = rotor copper loss = core losses = mechanical rotational losses. Determine the slip.

5.10 A four-pole 60-Hz three-phase wye-connected induction motor has a mechanical rotational loss of 500 W. At 5 percent slip, the motor delivers 30 hp at the shaft. Calculate (a) the rotor input, (b) the output torque, and (c) the developed torque.

5.11 A wound-rotor six-pole 60-Hz induction motor has a rotor resistance of 0.8 Ω and runs at 1150 rpm at a given load. The load on the motor is such that the torque remains constant at all speeds. How much resistance must be inserted in the rotor circuit to bring the motor speed down to 950 rpm? Neglect rotor leakage reactance.

5.12 A 400-V three-phase wye-connected induction motor has a stator impedance of $(0.6 + j1.2)\ \Omega$ per phase. The rotor impedance referred to the stator is $(0.5 + j1.3)\ \Omega$ per phase. Using the approximate equivalent circuit, determine the maximum electromagnetic power developed by the motor.

5.13 The motor of Problem 5.12 has a magnetizing reactance of 35 Ω. Neglecting the iron losses, at 3 percent slip calculate (a) the input current, and (b) the power factor of the motor. Use the approximate equivalent circuit.

5.14 On no-load a three-phase delta-connected induction motor takes 6.8 A and 390 W at 220 V. The stator resistance is 0.1 Ω per phase. The friction and windage loss is 120 W. Determine the values of the parameters X_m and R_m of the equivalent circuit of the motor.

5.15 On blocked-rotor, the motor of Problem 5.14 takes 30 A and 480 W at 36 V. Using the data of Problem 5.14, determine the complete exact equivalent circuit of the motor. Assume that the per phase stator and rotor leakage reactances are equal.

5.16 Determine the parameters of the approximate equivalent circuit of a three-phase induction motor from the following data:

No-load test:
  applied voltage = 440 V
  input current = 10 A
  input power = 7600 W

Blocked rotor test:
  applied voltage = 180 V
  input current = 40 A
  input power = 6240 W

The stator resistance between any two leads is 0.8 Ω and the no-load friction and windage loss is 420 W.

5.17 A four-pole 400-V three-phase 60-Hz induction motor takes 150 A of current at starting and 25 A while running at full-load. The starting torque is 1.8 times the torque at full-load at 400 V. If it is desired that the starting torque be the same as the full-load torque, determine (a) the applied voltage and (b) the corresponding line current.

5.18 An induction motor is started by a wye–delta switch. Determine the ratio of the starting torque to the full-load torque if the starting current is five times the full-load current and the full-load slip is 5 percent.

5.19 An induction motor is started at a reduced voltage. The starting current is not to exceed four times the full-load current, and the full-load torque is four times the starting torque. What is the full-load slip? Calculate the factor by which the motor terminal voltage must be reduced at starting.

5.20 The per phase parameters of the equivalent circuit of a double-cage rotor are $R_o = 0.4\ \Omega$, $X_o = 0.2\ \Omega$, $R_i = 0.04\ \Omega$, and $X_i = 0.8\ \Omega$. At starting, determine the ratio of the torques provided by the outer and inner cages.

5.21 At what slip will the torques contributed by the outer and inner cages of the rotor of Problem 5.20 be equal?

CHAPTER 6

Small AC Motors

6.1 INTRODUCTION

By small motors we imply that such motors generally have outputs of less than 1 hp (or 746 W), although in a special case the output may be greater than 1 hp. Small motors find applications in electric tools, appliances, and office equipment. We shall consider specific applications in a later section. Whereas permanent-magnet small dc commutator motors have numerous applications, such as in toys, control devices, and so on, in the following we consider only small ac motors.

Almost invariably, small ac motors are designed for single-phase operation. The three common types of small ac motors are:

1. Single-phase induction motors.
2. Synchronous motors.
3. Ac commutator motors.

Special types of small ac motors include:

1. Ac servomotors.
2. Tachometers.
3. Stepper motors.

235

6.2
SINGLE-PHASE INDUCTION MOTORS

Let us consider a three-phase cage-type induction motor (discussed in Chapter 5) running light. It may be experimentally verified that if one of the supply lines is disconnected, the motor will continue to run, although at a speed lower than the speed of a three-phase motor. Such an operation of a three-phase induction motor (with one line open) may be considered to be similar to that of a single-phase induction motor. Next, let the three-phase motor be at rest and supplied by a single-phase source. Under this condition, the motor will not start because we have a pulsating magnetic field in the air gap rather than a rotating magnetic field which is required for torque production, as discussed in Chapter 5. Thus we conclude that a single-phase induction motor is not self-starting, but will continue to run if started by some means. This implies that to make it self-starting, the motor must be provided with an auxiliary means of starting. Later, we shall examine the various methods of starting the single-phase induction motor.

Not considering the starting mechanism for the present, the essential difference between the three-phase and single-phase induction motor is that the latter has a single stator winding which produces an air gap field that is stationary in space, but alternating (or pulsating) in time. On the other hand, the stator of a three-phase induction motor has a three-phase winding that produces a time-invariant rotating magnetic field in the air gap. The rotor of the single-phase induction motor is generally a cage-type rotor and is similar to that of a three-phase induction motor. The rating of a single-phase motor of the same size as a three-phase motor would be smaller, as expected, and single-phase induction motors are most often rated as fractional horsepower motors.

Performance Analysis. The operating performance of the single-phase induction motor is studied on the basis of the following theories:

1. Cross-field theory.
2. Double-revolving field theory.

We will use the revolving-field theory in the following in analyzing the single-phase induction motor, although both theories yield identical results. It is interesting, however, to study qualitatively the mechanism of torque production in a single-phase induction motor, as viewed by cross-field theory. Figure 6.1(a) shows a single-phase induction motor, the stator winding of which carries a single-phase excitation and the rotor is at standstill. The stator mmf is shown as F_1. Since the rotor is stationary, it acts like a short-circuited transformer. As a consequence, the mmf due to the rotor currents simply opposes the stator mmf. The resulting field will be stationary in space but will pulsate in magnitude.

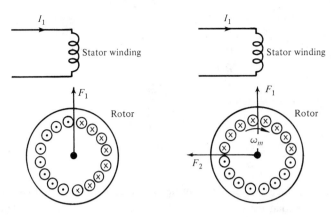

FIGURE 6-1. Representation of a single-phase induction motor: (a) standstill; (b) running.

Next, let the motor be started (by some means) and let it run at some speed, as shown in Fig. 6.1(b). Rotor conductors are now rotating in the magnetic field produced by the stator. Consequently, rotational voltages will be induced in the rotor conductors. Applying the right-hand rule (see Chapter 3) yields the directions of the voltages induced, and hence the current flow, in the rotor conductors, as given in Fig. 6.1(b). We now have two mmf's: F_1 due to the stator and F_2 due to the rotation of the rotor. These mmf's produce their respective airgap fields. These mmf's are displaced from each other in space, as shown in Fig. 6.1(b). To determine the time displacement between the two mmf's, notice that the rotor-induced voltage (because of rotation) is in time phase with the stator mmf. However, the rotor circuit being highly inductive, the rotor current lags the rotor voltage by almost 90°. Hence there is a time displacement (of about 90°) between the stator and rotor mmf's. Thus we fulfill the condition for the production of a rotating magnetic field (see Chapter 4), and the rotor continues to develop a torque as long as the rotor is running in the rotating magnetic field.

We see from the above that the magnetic field produced by the stator of a single-phase motor alternates through time. The field induces a current—and consequently, an mmf—in the rotor circuit and the resultant field rotates with the rotor. A single-phase induction motor may be analyzed by considering the mmf's, fluxes, induced voltages (both rotational and transformer), and currents that are separately produced by the stator and by the rotor. Such an approach leads to the cross-field theory. However, we can also analyze the single-phase motor in a manner similar to that for the polyphase induction motor. We recall from Chapter 5 that the polyphase induction motor operates on the basis of the existence of a rotating magnetic field. This approach is based on the concept that an alternating magnetic field is equivalent to two rotating magnetic fields rotating in opposite directions. When this concept is expressed mathematically, the alternating field is of the form

$$B(\theta,t) = B_m \cos \theta \sin \omega t \tag{6.1}$$

Then (6.1) may be rewritten as

$$B_m \cos \theta \sin \omega t = \frac{1}{2} B_m \sin (\omega t - \theta) + \frac{1}{2} B_m \sin (\omega t + \theta) \tag{6.2}$$

In (6.2), the first term on the right-hand side denotes a forward rotating field, whereas the second term corresponds to a backward rotating field. The theory based on such a resolution of an alternating field into two counter-rotating fields is known as the *double-revolving field theory*. The direction of rotation of the forward rotating field is assumed to be the same as the direction of the rotation of the rotor. Thus if the rotor runs at n rpm and n_s is the synchronous speed in rpm, the slip, s_f, of the rotor with respect to the forward rotating field is the same as s, defined by (5.1), or

$$s_f = s = \frac{n_s - n}{n_s} = 1 - \frac{n}{n_s} \tag{6.3}$$

But the slip, s_b, of the rotor with respect to the backward rotating flux is given by

$$s_b = \frac{n_s - (-n)}{n_s} = 1 + \frac{n}{n_s} = 2 - s \tag{6.4}$$

We know from the operation of polyphase motors that, for $n < n_s$, (6.3) corresponds to a motor operation and (6.4) denotes the braking region. Thus the two resulting torques have an opposite influence on the rotor.

The torque relationship for the polyphase induction motor is applicable to each of the two rotating fields of the single-phase motor. We notice from (6.2) that the amplitude of the rotating fields is one-half of the alternating flux. Thus the total magnetizing and leakage reactances of the motor can be divided equally so as to correspond to the forward and backward rotating fields. The approximate equivalent circuit of a single-phase induction motor, based on the double-revolving field theory, becomes as shown in Fig. 6.2(a). The torque–speed characteristics are qualitatively shown in Fig. 6.2(b). The following example illustrates the usefulness of the circuit.

EXAMPLE 6.1

For a 230-V one-phase induction motor, the parameters of the equivalent circuit, Fig. 6.2(a), are: $R_1 = R_2' = 8 \, \Omega$, $X_1 = X_2' = 12 \, \Omega$, and $X_m = 200 \, \Omega$. At a

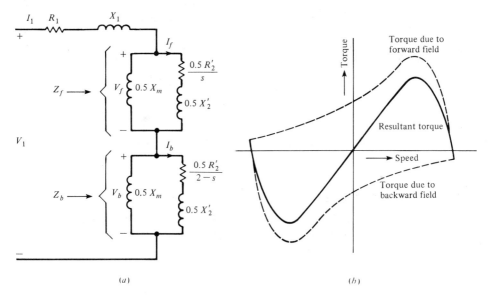

FIGURE 6-2. (a) Equivalent circuit and (b) torque-speed characteristics, based on double-revolving field theory of the single-phase induction motor.

slip of 4 percent, calculate (a) the input current, (b) the input power, (c) the developed power, and (d) the developed torque (at rated voltage). The motor speed is 1728 rpm.

SOLUTION
From Fig. 6.2(a):

$$\mathbf{Z}_f = \frac{(j100)\left(\dfrac{4}{0.04} + j6\right)}{j100 + \dfrac{4}{0.04} + j6} = 47 + j50 \ \Omega$$

$$\mathbf{Z}_b = \frac{(j100)\left(\dfrac{4}{1.96} + j6\right)}{j100 + \dfrac{4}{1.96} + j6} = 1.8 + j5.7 \ \Omega$$

$$\mathbf{Z}_1 = R_1 + jX_1 = 8 + j12 \ \Omega$$

$$\mathbf{Z}_{\text{total}} = 56.8 + j67.7 = 88.4\underline{/50°} \ \Omega$$

(a) $\quad\text{input current} \equiv I_1 = \dfrac{230}{88.4} = 2.6 \ \text{A}$

(b) $\quad\text{power factor} = \cos 50° = 0.64 \ \text{lagging}$

$\quad\text{input power} = (230)(2.6)(0.64) = 382.7 \ \text{W}$

(c)   Proceeding as in Example 5.2, we have

$$P_d = [I_1^2 \text{Re}(\mathbf{Z}_f)](1-s) + [I_1^2 \text{Re}(\mathbf{Z}_b)][1-(2-s)]$$
$$= I_1^2[\text{Re}(\mathbf{Z}_f) - \text{Re}(\mathbf{Z}_b)](1-s) = (2.6)^2(47-1.8)(1-0.04)$$
$$= 293.3 \text{ W}$$

(d) $$\text{torque} = \frac{P_d}{\omega_m} = \frac{293.3}{2\pi(1728)/60} = 1.62 \text{ N-m}$$ ∎

EXAMPLE 6.2

To reduce the numerical computation, Fig. 6.2(a) is modified by neglecting $0.5X_m$ in Z_b and taking the backward-circuit rotor resistance at low slips as $0.25R_2'$. With these approximations, repeat the calculations of Example 6.1 and compare the results.

SOLUTION

$$\mathbf{Z}_f = 47 + j50 \text{ } \Omega$$
$$\mathbf{Z}_b = 2 + j6 \text{ } \Omega$$
$$\mathbf{Z}_1 = 8 + j12 \text{ } \Omega$$
$$\mathbf{Z}_{\text{total}} = 57 + j68 = 88.7\underline{/50°} \text{ } \Omega$$

(a) $$I_1 = \frac{230}{88.7} = 2.6 \text{ A}$$

(b) $$\cos \varphi = 0.64 \text{ lagging}$$
$$\text{input power} = (230)(2.6)(0.64) = 382.7 \text{ W}$$

(c) $$P_d = (2.6)^2(47-2)(1-0.04) = 292.0 \text{ W}$$

(d) $$\text{torque} = \frac{292.0}{2\pi(1728)/60} = 1.61 \text{ N-m}$$ ∎

Following is a comparison of the results of the preceding two examples:

Example Number	Input Current (A)	Input Power (W)	Power Factor	Developed Torque (N-m)
6.1	2.6	382.7	0.64	1.62
6.2	2.6	382.7	0.64	1.61

This comparison indicates that the approximation suggested in Example 6.2 is adequate for most cases.

In the next example we show the procedure for efficiency calculations for a single-phase induction motor.

EXAMPLE 6.3

A one-phase 110-V 60-Hz four-pole induction motor has the following constants in the equivalent circuit, Fig. 6.2(a): $R_1 = R_2' = 2\ \Omega$, $X_1 = X_2' = 2\ \Omega$, and $X_m = 50\ \Omega$. There is a core loss of 25 W and a friction and windage loss of 10 W. For a 10 percent slip, calculate (a) the motor input current and (b) the efficiency.

SOLUTION

$$Z_f = \frac{(j25)\left(\frac{1}{0.1} + j1\right)}{j25 + \frac{1}{0.1} + j1} = 8 + j4\ \Omega$$

$$Z_b = \frac{(j25)\left(\frac{1}{1.9} + j1\right)}{j25 + \frac{1}{1.9} + j1} = 0.48 + j0.96\ \Omega$$

$$Z_1 = 2 + j2\ \Omega$$

$$Z_{\text{total}} = 10.48 + j6.96 = 12.6\underline{/33.6°}\ \Omega$$

(a) $$I_1 = \frac{110}{12.6} = 8.73\ \text{A}$$

(b) developed power = $(8.73)^2(8 - 0.48)(1 - 0.10) = 516$ W

output power = $516 - 25 - 10 = 481$ W

input power = $(110)(8.73)(\cos 33.6°) = 800$ W

$$\text{efficiency} = \frac{481}{800} = 60\%$$

■

Starting of Single-Phase Induction Motors. We already know that because of the absence of a rotating magnetic field, when the rotor of a single-phase induction motor is at standstill, it is not self-starting. The two methods of starting a single-phase motor are either to introduce commutator and brushes, such as in a repulsion motor (considered later), or to produce a rotating field by means of an auxiliary winding, such as by split phasing. We consider the latter method next.

From the theory of the polyphase induction motor, we know that in order to have a rotating magnetic field, we must have at least two mmf's which are displaced from each other in space and carry currents having different time phases. Thus, in a single-phase motor, a starting winding on the stator is provided as a source of the second mmf. The first mmf arises from the main stator winding. The various methods to achieve the time and space phase shifts between the main winding and starting winding mmf's are summarized below.

Split-Phase Motors. This type of motor is represented schematically in Fig. 6.3(a), where the main winding has a relatively low resistance and a high reactance. The starting winding, however, has a high resistance and a low reactance and has a centrifugal switch as shown. The phase angle α between the two currents I_m and I_s is about 30 to 45°, and the starting torque T_s is given by

$$T_s = K I_m I_s \sin \alpha \tag{6.5}$$

where K is a constant. When the rotor reaches a certain speed (about 75 percent of its final speed), the centrifugal switch comes into action and disconnects the starting winding from the circuit. The torque–speed characteristic of the split-phase motor is of the form shown in Fig. 6.3(b). Such motors find applications in fans, blowers, and so forth, and are rated up to $\frac{1}{2}$ hp.

A higher starting torque can be developed by a split-phase motor by inserting a series resistance in the starting winding. A somewhat similar effect may be obtained by inserting a series inductive reactance in the main winding. This reactance is short-circuited when the motor builds up speed.

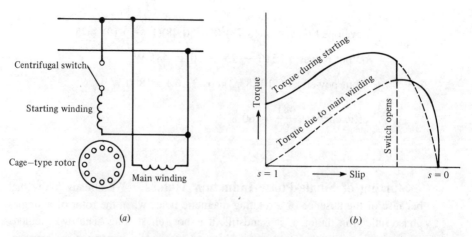

FIGURE 6-3. (a) Connections for a split-phase motor; (b) a torque-speed characteristic.

Capacitor-Start Motors. By connecting a capacitance in series with the starting winding, as shown in Fig. 6.4, the angle α in (6.5) can be increased. The motor will develop a higher starting torque by doing this. Such motors are not restricted merely to fractional-horsepower ratings, and may be rated up to 10 hp. At 110 V, a 1-hp motor requires a capacitance of about 400 µF, whereas 70 µF is sufficient for a $\frac{1}{8}$-hp motor. The capacitors generally used are inexpensive electrolytic types and can provide a starting torque that is almost four times that of the rated torque.

As shown in Fig. 6.4, the capacitor is merely an aid to starting and is disconnected by the centrifugal switch when the motor reaches a predetermined speed. However, some motors do not have the centrifugal switch. In such a motor, the starting winding and the capacitor are meant for permanent operation and the capacitors are much smaller. For example, a 110-V $\frac{1}{2}$-hp motor requires a 15 µF capacitance.

A third kind of the capacitor motor uses two capacitors: one that is left permanently in the circuit together with the starting winding, and one that gets disconnected by a centrifugal switch. Such motors are, in effect, unbalanced two-phase induction motors.

Shaded-Pole Motors. Another method of starting very small single-phase induction motors is to use a shading band on the poles, as shown in Fig. 6.5, where the main single-phase winding is also wound on the salient poles. The shading band is simply a short-circuited copper strap wound on a portion of the pole. Such a motor is known as the shaded-pole motor. The purpose of the shading band is to retard (in time) the portion of flux passing through it in relation to the flux coming out of the rest of the pole face. Thus the flux in the unshaded portion reaches its maximum before that located in the shaded portion. And we have a progressive shift of flux from the direction of the unshaded portion to shaded portion of the pole, as shown in Fig. 6.5. The effect of the

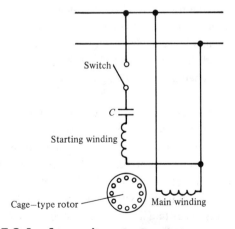

FIGURE 6-4. A capacitor-start motor.

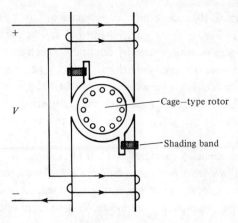

FIGURE 6-5. A shaded-pole motor.

progressive shift of flux is similar to that of a rotating flux, and because of it, the shading band provides a starting torque. Shaded-pole motors are the least expensive of the fractional horsepower motors and are generally rated up to $\frac{1}{20}$ hp.

6.3

SMALL SYNCHRONOUS MOTORS

The two common types of small synchronous motors are the *reluctance motor* and the *hysteresis motor*. These motors are constant-speed motors and are used in clocks, timers, turntables, and so forth.

Reluctance Motor. We are somewhat familiar with the reluctance motor from Chapter 4. We know that the torque in a reluctance motor is similar to the torque arising from saliency in a salient-pole synchronous motor. We may also recall that the time-average torque of a reluctance motor is nonzero only at one speed for a given frequency, and the power-angle characteristics of the motor are as discussed in Chapter 4. Schematically, a single-phase reluctance motor is shown in Fig. 6.6(a).

A reluctance motor starts as an induction motor, but normally operates as a synchronous motor. The stator of a reluctance motor is similar to that of an induction motor (single-phase or polyphase). Thus to start a single-phase motor, almost any of the methods discussed in Section 6.2 may be used. A three-phase reluctance motor is self-starting when started as an induction motor. After starting, to pull it into step and then to run it as a synchronous motor, a three-phase motor should have low rotor resistance. In addition, the combined inertia of the rotor and the load should be small. A typical construction of a four-pole rotor

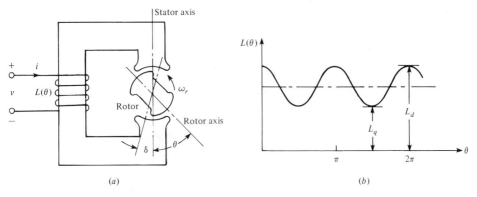

FIGURE 6-6. (a) A single-phase reluctance motor; (b) inductance of stator winding, as a function of rotor position.

is shown in Fig. 6.7. Here the aluminum in the slots and in spaces where teeth have been removed serves as the rotor of an induction motor for starting.

With L_d, the maximum inductance, and L_q, the minimum inductance, as defined in Fig. 6.6(b), the average value of the torque developed by the motor, for a single-phase current $i = I_m \sin \omega t$, is found to be

$$T_e = \tfrac{1}{8} I_m^2 (L_d - L_q) \sin 2\delta \tag{6.6}$$

where δ is the power angle (defined in Chapter 4).

For a three-phase reluctance motor having a rotor of the form shown in Fig. 6.7, the torque–speed characteristic takes the form shown in Fig. 6.8. The torques due to the induction motor action and due to reluctance are as labeled in Fig. 6.8. The power factor of a reluctance motor is poorer compared to that of an induction motor of similar rating.

EXAMPLE 6.4

A two-pole one-phase 60-Hz reluctance motor carries 5.66 A of current. The direct- and quadrature-axis inductances, L_d and L_q, respectively, are given by $L_d = 2L_q = 200$ mH. Determine (a) the rotor speed, (b) the power angle for maximum torque, and (c) the maximum value of the developed torque.

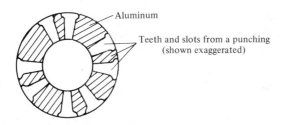

FIGURE 6-7. Rotor of a reluctance motor.

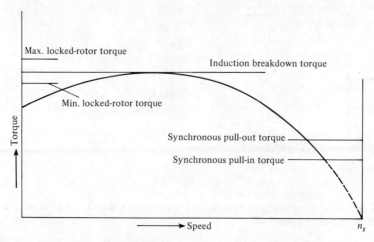

FIGURE 6-8. Torque-speed characteristic of a three-phase reluctance motor.

SOLUTION

(a) Synchronous speed:

$$n_s = \frac{120f}{p} = \frac{120 \times 60}{2} = 3600 \text{ rpm}$$

(b) For maximum torque:

$$\sin 2\delta = 1$$
$$2\delta = 90°$$
$$\delta = 45°$$

(c) The maximum torque for $\delta = 45°$ is obtained (6.6), which gives

$$T_e = \tfrac{1}{8}(\sqrt{2} \times 5.66)^2(200 - 100) \times 10^{-3} = 0.8 \text{ N-m} \qquad \blacksquare$$

Hysteresis Motor. Like the reluctance motor, a hysteresis motor does not have a dc excitation. Unlike the reluctance motor, however, the hysteresis motor does not have a salient rotor. Instead, the rotor of a hysteresis motor has a ring of special magnetic material, such as chrome, steel, or cobalt, mounted on a cylinder of aluminum or some other nonmagnetic material, as shown in Fig. 6.9. The stator of the motor is similar to that of an induction motor, and the hysteresis motor is started as an induction motor.

In order to understand the operation of the hysteresis motor, we may consider the hysteresis and eddy-current losses in the rotor. We observe that, as in an induction motor, the rotor has a certain equivalent resistance. The power dissipated in this resistance determines the electromagnetic torque developed by

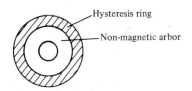

FIGURE 6-9. Rotor of a hysteresis motor.

the motor, as discussed in Chapter 5. We may conclude that the electromagnetic torque developed by a hysteresis motor has two components: one by virtue of the eddy-current loss and the other because of the hysteresis loss. We know that the eddy-current loss can be expressed as

$$p_e = K_e f_2^2 B^2 \tag{6.7}$$

where K_e is a constant, f_2 the frequency of the eddy currents, and B the flux density. In terms of the slip s, the rotor frequency f_2 is related to the stator frequency f_1 by

$$f_2 = s f_1 \tag{6.8}$$

Thus (6.7) and (6.8) yield

$$p_e = K_e s^2 f_1^2 B^2 \tag{6.9}$$

and the torque T_e is related to p_e by (see Chapter 5)

$$T_e = \frac{p_e}{s \omega_s} \tag{6.10}$$

so that (6.9) and (6.10) give

$$T_e = K' s \tag{6.11}$$

where $K' = K_e f_1^2 B^2 / \omega_s$ = a constant.

Next, for the hysteresis loss, P_h, we have

$$p_h = K_h f_2 B^{1.6} = K_h s f_1 B^{1.6} \tag{6.12}$$

and for the corresponding torque, T_h, we obtain

$$T_h = K'' \tag{6.13}$$

where $K'' = K_h f_1 B^{1.6} / \omega_s$ = a constant.

6.3 SMALL SYNCHRONOUS MOTORS

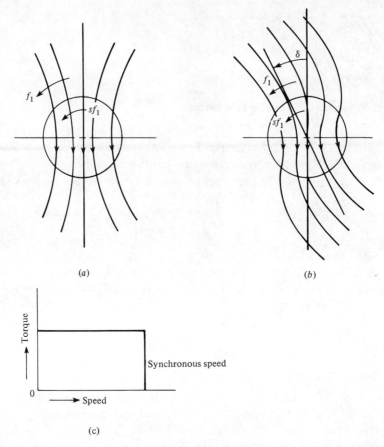

FIGURE 6-10. (a) Iron rotor, with no hysteresis in a magnetic field; (b) a rotor with hysteresis in a magnetic field; (c) torque characteristics of a hysteresis motor.

Notice that the component, T_e, as given by (6.11), is proportional to the slip and decreases as the rotor picks up speed. It is eventually zero at synchronous speed. This component of the torque aids in the starting of the motor. The second component, T_h, as given by (6.13), remains constant at all rotor speeds and is the only torque when the rotor achieves the synchronous speed. The physical basis of this torque is the hysteresis phenomenon, which causes a lag of the magnetic axis of the rotor behind that of the stator. In Fig. 6.10(a) and (b), respectively, the absence and the presence of hysteresis are shown measured by the shift of the rotor magnetic axis. The angle of lag δ, shown in Fig. 6.10(b), causes the torque arising from hysteresis. As mentioned above, this torque is independent of the rotor speed (shown in Fig. 6.10) until the breakdown torque.

6.4

AC COMMUTATOR MOTORS

In the preceding chapters we have distinguished a dc motor from an ac motor by the presence of a commutator in the dc motor. However, there exist a number of types of motors that have commutators but operate on ac. In the following discussion we consider only two types of ac commutator motors: the universal motor and the repulsion motor.

Universal Motors. A universal motor is a series motor that may be operated either on dc or on single-phase ac at approximately the same speed and power output, while supplied at the same voltage (on dc and on ac), and the frequency of the ac voltage does not exceed 60 Hz. Much of the application of universal motors is in domestic appliances and tools such as food mixers, sewing machines, vacuum cleaners, portable drills, and saws. There are two types of universal motors: *uncompensated* and *compensated*. The former is less expensive and simpler in construction. It is generally used for lower power outputs and higher speeds than the compensated motor. Typical torque–speed characteristics of the two types of motors are shown in Fig. 6.11. Notice that the compensated motor has better universal characteristics in that the operation on dc and on ac at 60 Hz is not substantially different in the high-speed range. The two types of motors differ in construction also. The uncompensated motor has salient poles and a concentrated field winding, whereas the field winding of a compensated type of motor is distributed on a nonsalient pole magnetic structure. In addition, some compensated-type universal motors have a compensating winding, and some compensated motors have only one field winding, which also serves as a compensating winding.

The principle of operation of the universal motor is similar to that of the dc series motor, the construction of the two being essentially similar. As seen from Chapter 3, the torque equation (3.10) of a dc motor may be written as

$$T_e = k_m \varphi I_a$$

where $k_m = Zp/2\pi a$ = a constant. If saturation is neglected, $\varphi = k_f I_f$, where k_f is a constant and I_f is the field current. Since $I_f = I_a = I$ in a series motor, denoting $k_f k_m = k$, the developed torque of a dc series motor is given as

$$T_e = kI^2 \qquad (6.14)$$

where k is a constant and I is the current through the field and the armature, the two being in series. Under ac operation, with a current $I_m \sin \omega t$, (6.14)

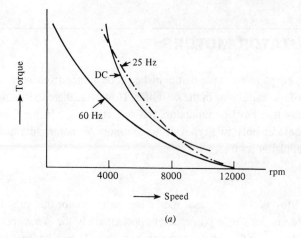

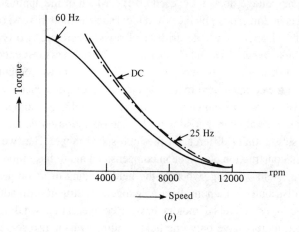

FIGURE 6-11. Characteristics of (a) uncompensated and (b) compensated universal motors.

yields

$$T_e = I_m^2 \sin^2 \omega t \tag{6.15}$$

Since $\sin^2 \omega t = \tfrac{1}{2}(1 - \cos 2\omega t)$, (6.15) becomes

$$T_e = kI^2(1 - \cos 2\omega t) \tag{6.16}$$

which is an expression for the instantaneous torque. The time-average value of T_e is, therefore,

$$(T_e)_{\text{average}} = kI^2 \tag{6.17}$$

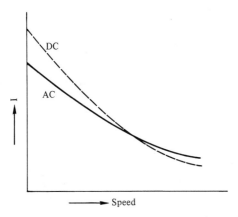

FIGURE 6-12. Current characteristics of a universal motor.

Hence the average torque is unidirectional and has the same magnitude as that with dc excitation, but the torque pulsates at twice the supply frequency.

In contrast to the solid poles of small dc series motors, the field structure of a universal motor is laminated to reduce hysteresis and eddy-current losses on ac operation. For a given voltage, the universal motor will draw less current (because of the reactance of the field and armature windings, see Fig. 6.12), develop less torque, and hence operate at a lower speed on ac compared to the operation of the motor on dc. Another difference between the ac and dc operations of a universal motor is that on ac the commutation is poorer because of the voltage induced, by transformer action, in the coils undergoing commutation.

A compensating winding is used to neutralize the reactance voltage of the armature winding. The compensating winding is connected in series with the armature, but is so arranged that the mmf of this winding opposes and neutralizes the armature mmf. Thus the compensating winding is displaced 90° (electrical) from the field winding, as shown in Fig. 6.13. The compensating winding not only neutralizes the armature reactance, but aids in commutation.

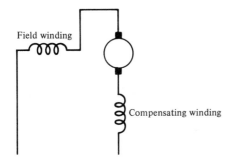

FIGURE 6-13. Conductively compensated universal motor.

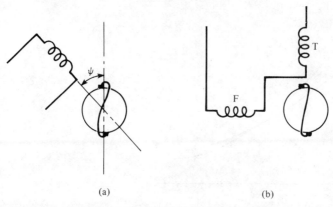

FIGURE 6-14. (a) A repulsion motor and (b) its equivalent.

Repulsion Motors. A repulsion motor is represented schematically in Fig. 6.14(a). The rotor is similar to that of a dc machine armature. However, the brushes are short-circuited along an axis displaced by an angle ψ from the axis of a single-phase stator winding. The stator winding is similar to that of a single-phase induction motor.

In order to understand the operation of the repulsion motor, we resolve the stator winding into two components on the stator as marked by F and T in Fig. 6.14(b). Then T may be considered as the primary of a transformer with the armature as the short-circuited secondary. When the stator is energized, the armature sets up its own mmf by induction action (via coupling through T). This mmf interacts with the stator mmf produced by F, resulting in a torque production. The current taken by the motor depends on the load. Hence the mmf due to F is directly proportional to the load current, and the motor has a variable-speed series motor characteristic. The motor is inherently compensated for armature reaction because an increase in the armature current results in an increase in the current through T and thereby increasing its mmf. This mmf is in opposition to the armature mmf. Inherent compensation is one of the advantages of a repulsion motor compared to a series motor. Low power factor and

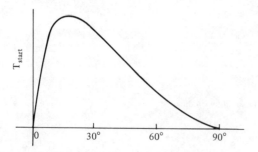

FIGURE 6-15. Starting torque vs. brush position.

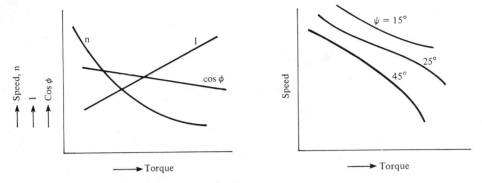

FIGURE 6-16. Characteristics of a repulsion motor.

tendency to spark at the brushes are among the disadvantages of a repulsion motor. These disadvantages can be offset by modified and compensated repulsion motors. But these motors are too specialized to be considered here.

Figures 6.15 and 6.16 show the various operating characteristics of a repulsion motor. The maximum starting torque is obtained when the brush shift is about 10 to 15°. At rated load, the optimum brush shift is between 15 and 25°.

6.5

TWO-PHASE MOTORS

Two-phase ac motors are usually used in instrumentation and control systems. Two such motors, discussed next, are ac tachometers and two-phase servomotors.

AC Tachometers. In many control applications it is desirable to express the speed of a motor as an ac voltage. A small two-phase induction motor, connected as shown in Fig. 6.17, is suitable for this purpose. The reference winding R is connected to an ac source of constant voltage and frequency. While the rotor is rotating, a voltage of the same frequency (as that of the reference voltage) is available at the control winding. The magnitude of this voltage is (ideally) linearly proportional to the speed of the rotor, and the phase of the voltage is fixed with respect to the reference voltage. In practice, the control winding is connected to a high-impedance amplifier and thus may be considered open-circuited for practical purposes. For a given current in the reference winding, $\mathbf{I}_r$, the control-winding voltage, $\mathbf{V}_c$, is given by

$$\mathbf{V}_c = k\mathbf{I}_r(\mathbf{Z}_f - \mathbf{Z}_b) \tag{6.18}$$

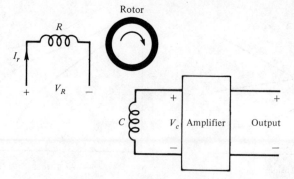

FIGURE 6-17. A two-phase tachometer.

where k is a constant and $\mathbf{Z}_f$ and $\mathbf{Z}_b$ are the forward and backward impedances of Fig. 6.2(a). Because $\mathbf{Z}_f$ and $\mathbf{Z}_b$ are functions of speed, $\mathbf{V}_c$ will measure the motor speed. Ac tachometers are commonly used in 400-Hz systems.

Two-Phase Servomotors. A two-phase servomotor, used in control applications, is very much similar to a two-phase ac tachometer just described. Servomotors are used to transform a time-varying signal to a time-varying motion: for example, in an antenna-positioning system. An arrangement of a two-phase servomotor is shown in Fig. 6.18. The motor has two stator field windings displaced from each other by 90 electrical degrees. In a conventional two-phase induction motor, the voltages V_r and V_c (Fig. 6.18) are equal and 90° displaced from each other in time. The speed of the induction motor depends on the load (or slip). However, the speed of the servomotor must be proportional to an input signal voltage. The control voltage V_c results from the amplitude of the voltage

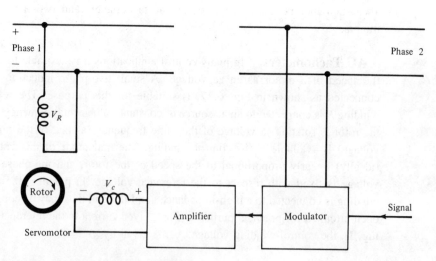

FIGURE 6-18. A two-phase servomotor.

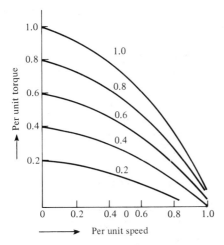

FIGURE 6-19. Speed–torque characteristics of a servomotor.

of phase 2, but modulated by the signal. The motor speed, in turn, will depend on the signal or control voltage, V_c. In this sense, the ac tachometer is a generator, whereas the servomotor functions as a motor. The torque–speed characteristics of a typical two-phase servomotor are shown in Fig. 6.19 for different values of the signal voltage.

6.6

STEPPER MOTORS

The stepper motor is not operated to run continuously. Rather, it is designed to rotate in steps in response to electrical pulses received at its input from a control unit. The motor indexes in precise angular increments. The average shaft speed of the motor, n_{average}, is given by

$$n_{\text{average}} = \frac{60 \text{ (pulses per second)}}{\text{number of phases in the winding}} \text{ rpm}$$

Two basic types of stepper motors are variable-reluctance (VR) and permanent-magnet (PM) stepper motors. In principle, a stepper motor is a synchronous motor, and typically has three-phase or four-phase windings on the stator. The number of poles on the rotor depends on the step size (or angular displacement) per input pulse. The rotor is either reluctance type or may be made of permanent magnet. When a pulse is given to one of the phases on the stator, the rotor tends to align with the mmf axis of the stator coil. These coils are sequentially switched and the rotor follows the stator mmf in sequence.

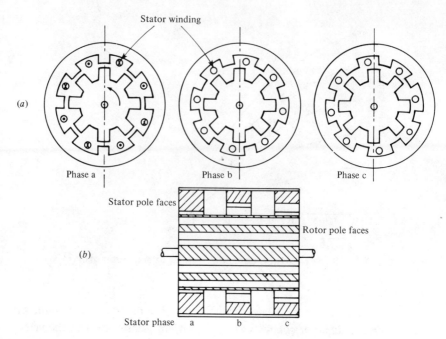

FIGURE 6-20. A variable-reluctance stepper motor: (a) phase 'a' energized and relative displacements of stator phases 'b' and 'c'; (b) cross-section of the assembled motor.

A variable-reluctance motor is shown in Fig. 6.20. The machine has a rotor with eight poles and three independent eight-pole stators arranged coaxially with the rotor as shown in Fig. 6.20(b). When phase a is energized, the motor poles align with the stator poles of phase a. Notice from Fig. 6.20(a) that the phase b stator is displaced from the phase a stator by 15° in the counterclockwise direction, and the phase c stator is further displaced from the phase b stator by another 15° in the counterclockwise direction. Now, when the current in phase a is turned off and phase b is energized, the rotor will rotate counterclockwise through 15°. Next, when phase b current is turned off and phase c is energized, the rotor will turn by another 15° in the counterclockwise direction. Finally, turning off the current in phase c and exciting phase a will complete one step (of 45°) in the counterclockwise direction. Additional current pulses in the sequence abc will produce further steps in the counterclockwise direction. Reversal of rotation is obtained by reversing the phase sequence to acb.

The stepping motion of the rotor of a stepper motor is typical of an undamped system, as illustrated in Fig. 6.21. The initial displacement of the rotor overshoots the final position and then gradually settles down to the final position. Some means of damping these oscillations is necessary in stepper motors. As the frequency of pulsing is increased, the period τ decreases. When τ is close to the oscillatory period t_1, the motor reaches its operating limit.

As seen from the preceding discussion, the step angle of a motor is determined

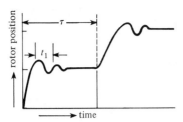

FIGURE 6-21. Undamped response of the rotor to step inputs.

by the number of poles. Typical step angles are 15°, 5°, 2°, and 0.72°. The choice of step angle depends on the angular resolution required for the application. The speed at which a stepping motor can operate is limited by the degree of damping existing in the system. Speeds up to 200 steps per second are typically attainable. A steady and continuous speed of rotation (*slewing*) greater than this value can be achieved, but the motor is then unable to stop the system in a single step.

A permanent-magnet stepper motor is shown in Fig. 6.22. The stator has a two-phase winding and the rotor has five pole pairs. For the position shown, phase *b* is energized and the rotor turns by (90° − 72°) = 18° to align with the pole of phase *b* winding. The switching of the phases for a clockwise rotation is shown in Fig. 6.23. The rotor achieves an equilibrium position after each step. If disturbed, the rotor will tend to return to its equilibrium position.

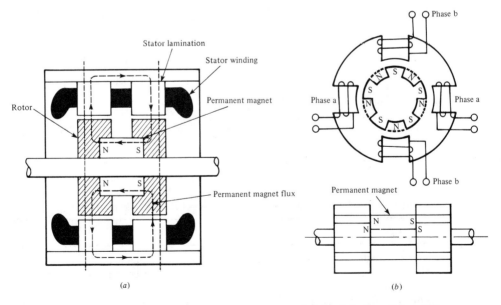

FIGURE 6-22. A permanent magnet stepper motor: (a) axial view; (b) radial cross-section.

6.6 STEPPER MOTORS

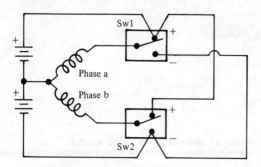

FIGURE 6-23. Circuit diagram and switching sequence for clockwise rotation of motor of Fig. 6.21.

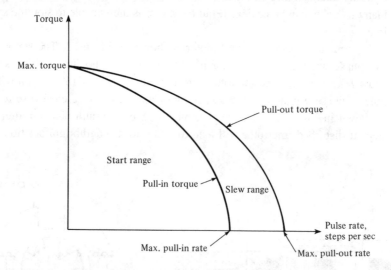

FIGURE 6-24. Torque–pulse rate characteristics.

TABLE 6.1 Comparison Between VR and PM Stepper Motors

VR Motor	PM Motor
No torque with no excitation	Residual torque with no excitation
Higher inductance and slower electrical response	Lower inductance and faster electrical response
Low rotor inertia and faster mechanical response	High inertia and slower mechanical response

The parameters that are important in the selection of a stepper motor are speed, torque, single-step response, holding torque, settling time, and slewing. In the slewing mode, the rotor does not come to a complete stop before the next pulse is switched on. The typical torque–steps per second characteristic of a step motor is shown in Fig. 6.24.

A comparison of variable-reluctance (VR) and permanent-magnet (PM) stepper motors is given in Table 6.1.

Stepper motors have a wide range of applications, including process control, machine tools, computer peripherals, and certain medical equipment.

QUESTIONS

6.1 List some of the commonly used ac fractional horsepower motors.

6.2 Why is it necessary to provide a means of starting a single-phase induction motor?

6.3 What are some of the commonly used methods to aid the starting of single-phase induction motors?

6.4 Explain why a single-phase induction motor will not self-start but will continue to run if started by an auxiliary means.

6.5 What is the significance of the shading band in a shaded-pole motor?

6.6 What are the two common types of single-phase synchronous motors?

6.7 Are single-phase synchronous motors self-starting?

6.8 List some of the applications of small synchronous motors.

6.9 Will a dc series motor work on ac?

6.10 What is a universal motor?

6.11 What is the purpose of compensating winding in a universal motor?

6.12 Explain the working of a repulsion motor.

6.13 Are ac commutator motors self-starting?

6.14 What is the difference between a two-phase ac tachometer and a two-phase ac servomotor?

6.15 What is a stepper motor?

6.16 Name the parameters that must be considered in selecting a stepper motor.

6.17 What is the difference between a variable-reluctance and a permanent-magnet stepper motor?

6.18 Name a few applications of stepper motors.

PROBLEMS

6.1 A 60-Hz single-phase induction motor has six poles and runs at 1000 rpm. Determine the slip with respect to (a) the forward rotating field and (b) the backward rotating field.

6.2 A 110-V one-phase two-pole 60-Hz induction motor is designed to run at 3420 rpm. The parameters of the motor equivalent circuit of Fig. 6.2(a) are $R_1 = R_2' = 6\,\Omega$, $X_1 = X_2' = 10\,\Omega$, and $X_m = 80\,\Omega$. Determine (a) the input current and (b) the developed torque at 3420 rpm.

6.3 The circuit of Fig. 6.2(a) is modified by neglecting $0.5X_m$ from the backward circuit of the rotor. In this circuit the rotor resistance is replaced by $0.3R_2'$. With these approximations, repeat the calculations of Problem 6.2 and compare the results.

6.4 If the motor of Problem 6.2 has a core loss of 8 W and a friction and windage loss of 10 W, determine the motor efficiency at 5 percent slip using (a) the equivalent of Fig. 6.2(a) and (b) the modified equivalent circuit of Problem 6.3.

6.5 What is the relative amplitude of the resultant forward-rotating flux density to the resultant backward-rotating flux density for the motor of Problem 6.2 at 5 percent slip?

6.6 Draw the modified equivalent circuit suggested in Problem 6.3. Determine the parameters of this modified circuit from the following test data:

    No-load test: input voltage, 110 V; input current, 2.5 A; input power, 45 W; friction and windage loss, 8 W
    Blocked-rotor test: input voltage, 46 V; input current, 1.8 A; stator resistance, 1.2 Ω (assume that $X_1 = X_2'$)

6.7 The direct- and quadrature-axis inductances of a reluctance motor are 60 mH and 25 mH. The motor operates at a power angle of 15° while taking 2.2 A of current. Calculate the torque developed by the motor.

6.8 Determine the torque developed by the motor of Problem 6.7 if the motor is supplied by a 110-V 60-Hz source and the power angle is 15°. Neglect the motor winding resistance.

6.9 What is the operating speed of a reluctance motor if it has four poles and is operated at 400 Hz?

6.10 The hysteresis loss in the material of a two-pole 60-Hz hysteresis motor is 600 W. Determine the torque developed by the motor. If 8 W is lost because of various rotational losses, calculate the power and the torque available at the shaft.

6.11 A 110-V universal motor has an input impedance of $(30 + j40)$ Ω. The armature has 36 conductors and is lap-wound. The field has two poles and the flux per pole is 30 mWb. Determine the torque developed by the motor.

6.12 The motor of Problem 6.11 has a total no-load and rotational loss of 30 W at 4000 rpm. Calculate (a) the power factor and (b) the efficiency of the motor.

CHAPTER 7

Electric Drives

7.1 GENERAL REMARKS

In Chapters 3, 4, and 5 we have discussed the principles of operations of dc, induction, and synchronous motors, respectively. In those chapters we have also discussed the possible means of controlling some of the motor characteristics such as speed, torque, and so on. However, we did not get into the details of any present-day motor controllers. There are numerous motor applications which require precise control of one or more of the motor output parameters, such as speed, torque, and output power. The significance of the problem may be understood by considering the simple example of a variable-discharge pump. We may either choose a nonadjustable drive for the pump and control the flow by throttling, or use an adjustable drive and no throttling. The results of the two choices in terms of power requirements are shown in Fig. 7.1. From 7.1 it is clear that it is better to choose an adjustable-speed drive that can rotate the pump at the correct speed to deliver the exact flow and pressure required. We see that an adjustable-speed system (in this example, a hydroviscous drive) has approximately 57 percent less wasted power than a throttling system when the pump flow is reduced to 20 percent of maximum pump flow. Figure 7.1 also shows that even at 70 percent of maximum pump flow, the throttling system wastes

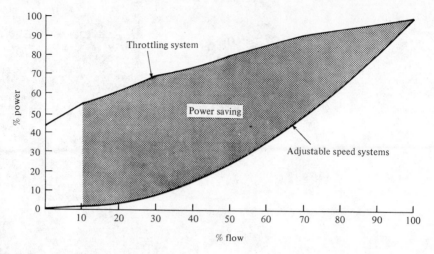

FIGURE 7-1. Comparison of % power required for % flow, hydroviscous adjustable-speed drive vs. throttling system.

42 percent more power than the adjustable-speed drive. These figures indicate that the power saved by an adjustable-speed drive would pay for the drive in a few months. Adjustable-speed drives allow the speed of the load to be varied continuously to meet the demands of the process and avoid or reduce the waste of energy that occurs otherwise.

The preceding is a simple, yet illustrative example of where the use of an adjustable-speed drive is most desirable. Depending on the application, there are several drive systems available and the selection depends on a large number of factors, such as initial cost, efficiency, reliability, and maintenance. Some of the applications requiring adjustable drives and motor control are in the glass, food, machine tool, material handling, petrochemical, water and wastewater treatment, paper and paper converting, test stand, and textile and synthetic fiber industries. The range of applications of adjustable drives is thus very broad.

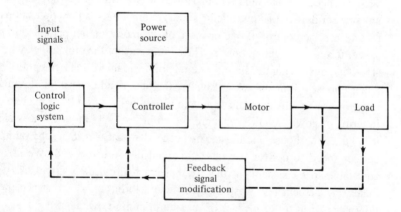

FIGURE 7-2. Block diagram representation of a motor control system.

A basic scheme for motor control is shown in Fig. 7.2. This figure illustrates a total motor system, including load and power source. The feedback loops are shown by dashed lines, since many motor control schemes are "open-loop." The primary concern of this chapter will be with the box labeled "Controller." Because much of the present-day motor control is accomplished by solid-state electronic devices, we begin with a brief review of the devices used in power electronics and motor control.

7.2 POWER SEMICONDUCTORS

Many types of semiconductors are available for electronic motor-control applications. The type to be used in a specific application will depend primarily on the power, voltage, and current requirements of the motor to be controlled. Other factors include the ambient temperature, the control modes to be used, and overall system cost considerations.

The power semiconductors and their associated circuitry are often referred to as the *power circuit* of a motor control system. The controller box in Fig. 7.2 contains the power circuit. There is another grouping of circuits and components required in a complete motor-control system, which operates at very low levels of voltage and current as compared to the power circuit. This group is referred to as the *control logic* or *gate-control circuitry*. Its function is to direct the operation of the motor, and it includes circuitry for turning the power semiconductors on and off.

Table 7.1 lists the principal semiconductor devices used in motor control systems, together with the standard symbols for these devices in circuit diagrams and the "state-of-the-art" maximum voltage, current, and time-response capabilities of each class of device. The time or speed parameter in Table 7.1 generally refers to the minimum turnoff time. Although the zener diode is not a true power-controlling device, it is included in Table 7.1 because it is widely used as a voltage control and sensing device in many motor controllers. Of the devices listed in Table 7.1, only the silicon rectifier, silicon-controlled rectifier (SCR) or thyristor, and power transistor are considered in detail here.

Silicon Rectifier. Symbolic representations and the ideal vi characteristic of a silicon rectifier are given in Fig. 7.3, where it is assumed that the diode acts like an ideal switch. In practice, however, the silicon rectifier is not a perfect conductor, and has a forward voltage drop of approximately 1 V at all currents within the rating of the rectifier.

The principal parameters of a silicon rectifier are the repetitive peak reverse voltage (PRV) or blocking voltage, average forward current, and maximum operating junction temperature (which is approximately 125°C for most silicon

TABLE 7.1 Power Semiconductors

Power Semiconductor	Symbol[a]	Maximum Capabilities		
		Voltage (V)	Rms Current (A)	Speed (μs)
Silicon rectifier		5000	7500	—
Silicon-controlled rectifier (SCR)		5000	3000	1
Bidirectional switch (TRIAC)		1000	2000	1
Gate turn-off SCR (GTO) Gate-controlled switch (GCS)		400 1000	— 200	0.2 2
Power transistor		3000	500	0.2
Darlington		1000	200	1
Zener diode		500	—	100

[a] A, anode; K, cathode; G, gate; E, emitter; C, collector; B, base.

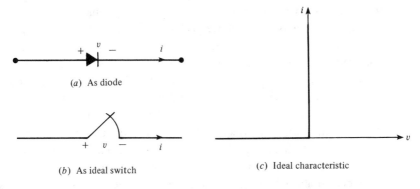

FIGURE 7-3. An ideal diode and its *vi*-characteristic.

devices). Another important characteristic in certain applications is the recovery time of the rectifiers. The recovery time is the time following the end of forward conduction before the rectifier assumes its full reverse blocking capability.

In motor control systems, silicon rectifiers commonly function as "freewheeling" diodes, providing a path for continuation of motor current following the switching off of another power device between the motor terminals and the power source.

Silicon-Controlled Rectifier (SCR) or Thyristor. This device is also frequently termed a *thyristor*. A symbolic representation of an SCR and its ideal *vi* characteristic are shown in Fig. 7.4. The SCR has three terminals: anode, cathode, and gate. A gate signal is used to turn on the SCR. In the reverse, or blocking, direction, the SCR functions very much like the silicon rectifier. In the forward direction, conduction can be controlled to a limited extent by the action of the gate, which is connected to low-signal circuitry that can be electrically isolated from the power circuitry connected to the anode and cathode. Control is limited, and only momentary, because the gate can turn on the device (that is, initiate the conditions for forward current flow) but cannot stop the current flow. Thus, when the current begins to flow through the anode, the SCR latches on and remains in the conductive state until turned off by external means,

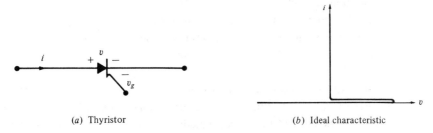

FIGURE 7-4. An ideal SCR and its *vi*-characteristic.

7.2 POWER SEMICONDUCTORS

in a process known as *commutation*. The principal parameters for applying SCRs to motor controllers are repetitive peak reverse voltage (PRV); maximum value of average on-state current, which determines the heating within the SCR; maximum value of on-state current, which is the rating of the metal conductor portions of the device; peak one-cycle on-state current, which is the surge current limit; the critical rate of rise of forward blocking voltage, which has two values—initial (when the SCR is first turned on) and reapplied (following commutation) turn-off time, which is the off-time required following commutation before forward voltage can be reapplied; the maximum rate of rise of anode current during turn-on; and maximum operating junction temperature. Although not included above, thermal management of the SCR—design of its heat sink, mounting method, and cooling—is extremely critical in all applications.

For motor controller use, there are two broad classes of SCRs. *Inverter-type* SCRs are applicable in inverters, cycloconverters, and brushless dc motor systems; *chopper-type* SCRs apply in choppers, phase-controlled rectifiers, regulators, and the like. The primary difference between the two types is in respect to time of response. Inverter types are generally more costly than chopper types. There are further ways to classify SCRs, especially in the smaller ratings used in control and communications applications. Many SCRs rated 35 A (rms) or below can be packaged in plastic cases, which makes them cheaper than metallic SCRs of similar ratings. Another classification relates to the gate signal. A very useful class of SCRs for position sensing in motor control systems is the light-activated class, or LASCR. In place of energy injection by electrical current, the LASCR is triggered on by photon energy. Light activation is possible in all the other three- and four-terminal devices listed in Table 7.1.

The forward voltage drop of an SCR varies considerably during anode current conduction and may be very large during the first few microseconds after turn-on. The average value is 1.5 to 2.0 V.

Power Transistor. When used in motor control circuits, power transistors are almost always operated in a switching mode. The transistor is driven into saturation and the linear gain characteristics are not used. The common-emitter configuration is the most common, because of the high power gain in this connection. The collector–emitter saturation voltage, $V_{CE(SAT)}$ for typical power transistors is from 0.2 to 0.8 V. This range is considerably lower than the on-state anode-to-cathode voltage drop of an SCR. Therefore, the average power loss in a power transistor is lower than that in an SCR of equivalent power rating. The switching times of power transistors are also generally faster than those of SCRs, and the problems associated with turning off or commutating an SCR are almost nonexistent in transistors. However, a power transistor is more expensive than an SCR of equivalent power capability. In addition, the voltage and current ratings of power transistors available are much lower than those of existing SCRs. It has already been stated that the maximum ratings listed in Table 7.1 are generally unobtainable concurrently in a single device.

This is particularly true of power transistors. Devices with voltage ratings of 1000 V or above have limited current ratings of 10 A or less. Similarly, the devices with higher current ratings, 50 A and above, have voltage ratings of 200 V or less. There has been relatively little operating experience in practical motor-control circuits with transistors whose voltage or current capabilities approach the upper limits listed in Table 7.1. For handling motor control requiring large current ratings at 200 V or below, it has been common to parallel transistors of lower current rating. This requires great care to assure equal sharing of collector currents and proper synchronization of base currents among the paralleled devices.

Typical specifications for power switching transistors of significance for motor control application include (Fig. 7.5) breakdown voltage, specified by the symbols BV_{CEO}—collector-to-emitter breakdown voltage with the base open—and BV_{CBO}—collector-to-base breakdown voltage with emitter open; collector saturation voltage, $V_{CE(SAT)}$; emitter–base voltage rating, V_{EBO}; maximum collector current, I_c, average and peak; forward current transfer ratio, H_{FE}, the ratio of collector to base current in the linear region; power dissipation; maximum junction temperature, typically 150 to 180°C; switching times: rise time; storage time; and fall time. Sometimes these switching times are related to a maximum frequency of switching.

The thermal impedances and temperature coefficients are also important parameters. In paralleling power transistors, the variation of device characteristics with temperature becomes especially significant. The I_c–V_{BE} characteristic is extremely temperature sensitive.

Figure 7.6 shows a family of power semiconductors: rectifiers, SCRs, and transistors. Disk type (or hockey puck) devices—one mounted and with heat sink—are shown in Fig. 7.7. Heat sinks and various components used in power

Maximum Allowable Ratings

Types	Repetitive peak off-state voltage, V_{DRM}[a] $T_j = 40$ to $+125°C$	Repetitive peak reverse voltage, V_{RRM}[a] $T_j = 0.40$ to $+125°C$	Nonrepetitive peak reverse voltage, V_{RSM}[a] $T_j = +125°C$
C449PN	1800 volts	1800 volts	2040 volts
C449PS	1700	1700	1920
C449PM	1600	1600	1790
C449PE	1500	1500	1700

[a] Half sinewave waveform, 10 ms max pulse width. Consult factory for lower-rated voltage devices.

Peak one cycle surge (nonrepetitive) on-state current, I_{TSM}	6500 amperes
Critical rate-of-rise of on-state current, Nonrepetitive	500 A/μs
Critical rate-of-rise of on-state current, repetitive	300 A/μs
Average gate power dissipation, $P_{G(av)}$	5 watts
Storage temperature, T_{stg}	−40 to +150°C
Operating temperature, T_j	−40 to +125°C
Mounting force required	3000 lb + 500 lb − 0 lb
	13.3 KN + 2.2 KN − 0 KN

FIGURE 7-5. SCR data sheet. (Courtesy General Electric Company)

Test	Symbol	Min.	Types	Max.	Units	Test conditions
Repetitive peak reverse and on-state current	I_{RRM} and I_{DRM}	–	10	25	mA	$T_j = +25°C, V = V_{DRM} = V_{RRM}$
Repetitive peak reverse and off-state blocking Current	I_{RRM} and I_{DRM}	–	45	60	mA	$T_j = +125°C, V = V_{DRM} = V_{RRM}$
Thermal resistance	$R_{\theta JC}$	–	–	.04	°C/watt	Junction-to-case – Double-side cooled
Critical linear rate-of-Rise of off-state voltage (Higher values may cause device switching)	dv/dt	200	–	–	V/μsec	$T_j = +125°C, V_{DRM} = 0.80$ rated, gate -open. Exponential or linear rising wave form. Exponential $di/dt = 0.8\, V_{DRM}\,(0.632)/\tau$
			Higher minimum dv/dt selections available – consult factory			
Gate trigger current	I_{GT}	–	–	200	mA dc	$T_C = +25°C, V_D = 6$ V dc, $R_L = 3$ ohms
		–	–	150		$T_C = +125°C, V_D = 6$ V dc, $R_L = 3$ ohms
Gate trigger voltage	V_{GT}	–	–	3	V dc	$T_C = 25$ to $+125°C, V_D = 6$ V dc $R_L = 3$ ohms
		–	–	5,...,3		$T_C = -40$ to $25°C, V_D = 6$ V dc, $R_L = 3$ ohms
Peak on-state voltage	V_{TM}	–	–	2.8	volts	$T_C = 25°C, I_T = 2000$ A peak Duty cycle ≤ 0.01%
Conventional circuit Commutated turnoff Time (with reverse voltage) C449-60 C449-60	t_q	– –		60 40	μsec	(1) $T_C = +125°C$ (2) $I_{TM} = 500$ A (3) $V_R \geq 50$ volts (4) 80% of V_{DRM} reapplied (5) Rate-of-rise of off-state voltage = 200 V/μs (6) Gate bias = open during turnoff Interval, 0 Volts, 100 Ohms (7) Duty cycle ≤ .01%
Conventional circuit Commutated turnoff Time (with feedback diode) C449-60 C449-40	t_q	– –		60 b 40 b	μsec	(1) $T_C = +125°C$ (2) $I_{TM} = 500$ A (3) $V_R = 2$ volts, minimum (4) 80% of V_{DRM} reapplied (5) Rate-of-rise of off-state voltage = 200 V/μs (6) Gate bias = open during turnoff interval (7) Duty cycle ≤ 0.01%

b Consult factory for maximum turn-off time.

FIGURE 7-5. (Continued)

electronics are shown in Fig. 7.8. Having acquired a knowledge of power semiconductor devices, we are now ready to study their applications to electric drives.

7.3

CONTROLLERS FOR DC MOTORS

The torque and speed characteristics (under steady state) of dc motors are essentially governed by the equations derived in Chapter 3. For convenience we repeat these pertinent equations:

$$E = \frac{\varphi n Z p}{60 a} \tag{7.1}$$

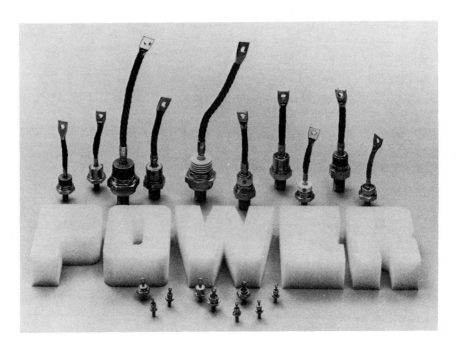

FIGURE 7-6. A family of power semiconductors—general purpose diodes. (Courtesy of Westinghouse Electric Corporation)

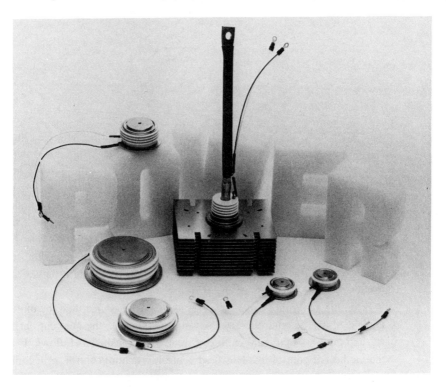

FIGURE 7-7. Phase control thyristors and mounting with heat sink. (Courtesy of Westinghouse Electric Corporation)

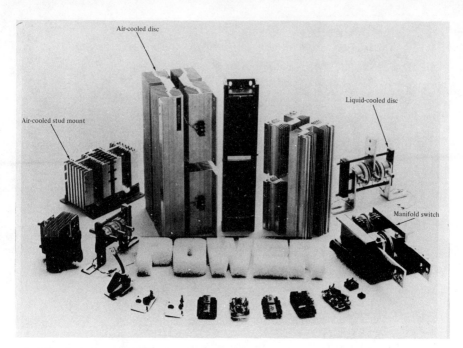

FIGURE 7-8. Heat sinks and other components for power semiconductor circuits. (Courtesy of Westinghouse Electric Corporation)

$$T_e = \frac{Zp}{2\pi a} \varphi I_a \tag{7.2}$$

$$\omega_m = \frac{V_t - I_a R_a}{k_1 \varphi} \tag{7.3}$$

The above are, respectively, (3.7), (3.10), and (3.12), where E is the voltage induced in the armature, volts; I_a the armature current, amperes; φ the flux per pole, webers; Z the number of armature conductors; a the number of parallel paths; p the number of poles; ω_m the armature speed, radians per second; n the armature speed, revolutions per minute; V_t the armature terminal voltage, volts; and $k_1 = Zp/2\pi a$, a constant; and these equations indicate the great flexibility of controlling the motor. For instance, the speed of a dc motor may be varied by varying V_t, R_a, or φ (that is, the field current). Control of dc motors is governed by (7.1) through (7.3), and the various practical schemes are manifestations of these equations in one form or another.

From the governing equations (7.2) and (7.3), it is clear that the motor torque and speed can be controlled by controlling φ (that is, the field current), V_t, and R_a, and changes in these quantities can be accomplished as follows. In essence, the method of control involves field control, armature control, or a combination of the two.

Chopper Control. Resistance control, either in the field or armature of a dc motor, results in a poor efficiency. High-power solid-state controllers offer the most practical, reliable, and efficient method of motor control. The most commonly used solid-state devices in motor control are power transistors and thyristors (or SCRs). The principal differences between the two are that the transistor requires a continuous driving signal during conduction, whereas the thyristor requires only a pulse to initiate conduction; and the transistor switches off when the driving signal is removed, but the thyristor turns off when the load current is reduced to zero or when a reverse-polarity voltage is applied.

Utilizing power semiconductors, the dc chopper is the most common electronic controller used in electric drives. In principle, a chopper is an on-off switch connecting the load to and disconnecting it from the battery (or dc source), thus producing a chopped voltage across the load. Symbolically, a chopper as a switch is represented in Fig. 7.9(a), and a basic chopper circuit is shown in Fig. 7.9(b). In the circuit shown in Fig. 7.9(b), when the thyristor does not conduct, the load current flows through the freewheeling diode D. From Fig. 7.9 it is clear that the average voltage across the load, V_0, is given by

$$V_0 = \frac{t_{on}}{t_{on} + t_{off}} V = \frac{t_{on}}{T} V_b = \alpha V \tag{7.4}$$

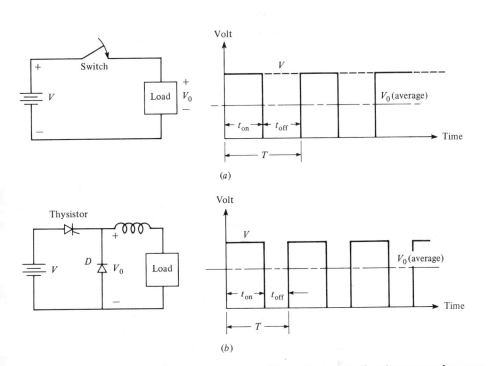

FIGURE 7-9. (a) Symbolic representation; (b) basic circuit of a chopper and output waveforms.

where the various times are shown in Fig. 7.9, T is known as the chopping period, and $\alpha = t_{on}/T$ is called the *duty cycle*. Thus the voltage across the load varies with the duty cycle.

There are three ways in which the chopper output voltage can be varied, and these are illustrated in Fig. 7.10. In the first method, the chopping frequency is kept constant and the pulse width (or, on-time t_{on}) is varied, and the method is known as *pulse-width modulation*. The second method, called *frequency modulation*, has either t_{on} or t_{off} fixed, and a variable chopping period, as indicated in Fig. 7.10(b). The preceding two methods can be combined to obtain pulse-width and frequency modulation, shown in Fig. 7.10(c), which is used in current limit control. In a method involving frequency modulation, the frequency must not be decreased to a value that may cause a pulsating effect or a discontinuous armature current; and the frequency should not be increased to such a high value so as to result in excessive switching losses. The switching frequency of most

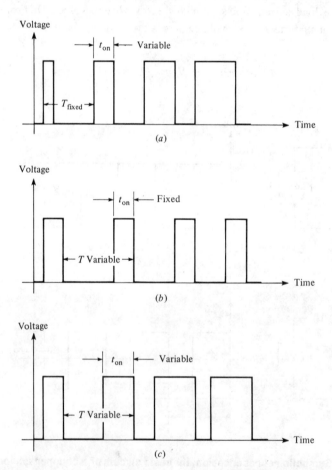

FIGURE 7-10. (a) Constant frequency, variable pulsewidth; (b) variable frequency, constant pulsewidth; (c) variable frequency, variable pulsewidth.

choppers for electric drives range from 100 to 1000 pulses per second. The drawback of a high-frequency chopper is that the current interruption generates a high-frequency noise.

A chopper circuit in a simplified form is shown in Fig. 7.11(a), where the chopper is shown to supply a dc series motor. The circuit is a pulse-width modulation circuit, where t_{on} and t_{off} determine the average voltage across the motor, as given by (7.4). As mentioned in Section 7.3, the SCR cannot turn itself off once it begins to conduct. So to turn the SCR off, we require a commutating circuit that impresses a negative voltage on the SCR for a very short time (of the order microseconds). The circuitry for commutation is often

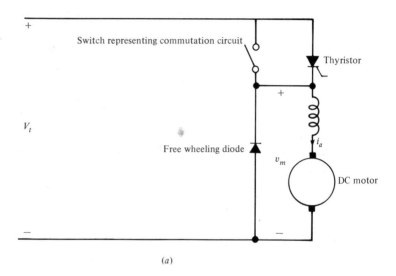

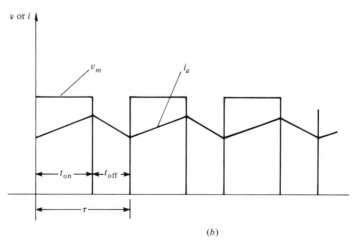

FIGURE 7-11. (a) A dc motor driven by a chopper; (b) motor voltage and current waveforms.

7.3 CONTROLLERS FOR DC MOTORS

quite involved. For our purposes, we denote the commutating circuit by a switch in Fig. 7.11(a).

The motor current and voltage waveforms are shown in Fig. 7.11(b). The SCR is turned on by a gating signal at $t = 0$. The armature current, i_a, builds up and the motor starts and picks up speed. After the time t_{on}, the SCR is turned off, and remains off for a period t_{off}. During this period, the armature current continues to drop through the freewheeling diode circuit. Again, at the end of t_{off}, the SCR is turned on and the on-off cycle continues. The chopper thus acts as a variable-voltage source.

Figure 7.12 shows the Jones chopper circuit, which works as follows: S_1 is turned on, resulting in current flow through S_1-L_2-motor; L_2 and L_1 are magnetically coupled (both windings are usually wound on the same magnetic core). Therefore, current flow through L_2 causes a proportional current through L_1-D_1-C, charging C, with the lower plate positive. The pulse through S_1 is ended by turning on S_2, which reverse-biases S_1 and causes the typical sinusoidal pulse through the path C-S_2-L_2-motor. Shortly after this current pulse reaches its maximum value, D_2 is forward-biased and begins to conduct and S_2 is reverse-biased and turns off. During the next period of operation, beginning with the turning on of S_1, C reverses its voltage through the path C-S_1-L_1-D_1, which also contributes to the load current through L_2. When the voltage on C is reversed, further charging may continue through D_1 by the coupling action of load current in L_2 until D_1 is reversed-biased. When D_1 is turned off by this reverse-bias condition and C is charged, with the lower plate positive, the circuit is in the condition to repeat the sequence of events described previously. The disadvantages of the Jones circuit are the size and weight of the coupled inductances, L_1 and L_2. However, this circuit has resulted in low manufacturing costs, partly

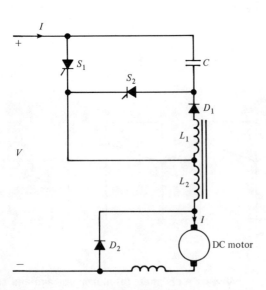

FIGURE 7-12. Jones chopper circuit.

because of the use of a commutating capacitor of lower size. Other advantages of the Jones chopper circuit are: (1) the circuit can be utilized for shunt or series motors; and (2) the output voltage may be varied either by varying the chopper frequency or pulsewidth.

The equations governing the design of the Jones chopper are as follows:

$$t_{\text{off}} = \frac{CV}{I} \tag{7.5}$$

$$\tfrac{1}{2}CV_c^2 = \tfrac{1}{2}L_1 I^2 \tag{7.6}$$

When S_1 is turned on, (7.6) implies that the energy stored in the capacitance is transferred to the inductance L_1. In (7.5) and (7.6), I is the rated current of the circuit and V_c is the initial charge across the capacitance. The optimum frequency for a minimum commutation loss and smaller commutation loss requires that

$$C \geq \frac{\pi I t_{\text{off}}}{2V} \tag{7.7}$$

and

$$L_1 = L_2 \tag{7.8}$$

where the symbols are as defined in Fig. 7.12.

EXAMPLE 7.1

For a dc series motor driven by a chopper, the following data are given: supply voltage, 440 V; duty cycle, 30 percent; motor circuit inductance, 0.04 H; and maximum allowable change in the armature current, 8 A. Determine the chopper frequency.

SOLUTION
From (7.4), the average voltage across the motor is

$$V_0 = 0.3 \times 440 = 132 \text{ V}$$

The voltage across the motor circuit inductance is

$$V_L = 440 - 132 = 308 \text{ V}$$

This voltage is also given by

$$V_L = L \frac{\Delta I}{\Delta t}$$

where $V_L = 308$ V, $L = 0.04$ H, $\Delta I = 8$ A, and $\Delta t = t_{on}$. Hence

$$t_{on} = \frac{0.04 \times 8}{308} = 1.04 \text{ ms}$$

But

$$\alpha = 0.3 = \frac{t_{on}}{t_{on} + t_{off}} = \frac{t_{on}}{T}$$

or

$$T = \frac{1.04}{0.3} = 3.46 \text{ ms}$$

Hence

$$\text{frequency} = \frac{1}{T} = 289 \text{ pulses/s}$$

■

EXAMPLE 7.2

If the chopper of Example 7.1 is to be designed for a current rating of 50 A, determine the values of C, L_1, and L_2 of the circuit shown in Fig. 7.12.

SOLUTION

From Example 7.1,

$$t_{off} = T - t_{on} = (3.46 - 1.04) = 2.42 \text{ ms}$$

From (7.7) we have

$$C \geq \frac{\pi I}{2V} t_{off}$$

$$\geq \frac{\pi \times 50}{2 \times 440} \times 2.42 \times 10^{-3} \geq 432 \text{ μF}$$

or

$$C = 450 \text{ μF}$$

Now

$$t_{off} = \sqrt{L_1 C}$$

or

$$L_1 = \frac{t_{\text{off}}^2}{C} = \frac{(2.42)^2 \times 10^{-6}}{450 \times 10^{-6}} = 13 \text{ mH}$$

and

$$L_1 = L_2 = 13 \text{ mH} \qquad \blacksquare$$

Converters. A converter changes an ac input voltage to a controllable dc voltage. Converters have an advantage over choppers in that *natural* or *line commutation* is possible in converters, and therefore no complex commutation circuitry is required. Controlled converters use SCRs and operate either on single-phase or three-phase ac. The four types of phase-controlled converters commonly used for dc motor control are:

1. Half-wave converters.
2. Semiconverters.
3. Full converters.
4. Dual converters.

Each of the above could be either a single-phase or a three-phase converter. Semiconverters are one-quadrant converters in that the polarities of voltage and current at the dc terminals do not reverse. In a full converter, the polarity of the voltage reverses, but the current is unidirectional. In this sense, a full converter is a two-quadrant converter. Dual converters are four-quadrant converters.

A half-wave converter and its quadrant operation are shown in Fig. 7.13. This type of converter is used for motors of ratings up to $\frac{1}{2}$ hp. Other types of single-phase converters have been used in drives rated up to 100 hp. Single-phase semi-, full, and dual converters and their respective quadrant operations are shown in Fig. 7.14(a)–(c), respectively.

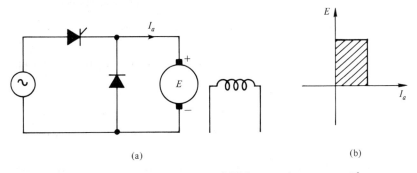

FIGURE 7-13. (a) A half-wave converter; and (b) its quadrant operation.

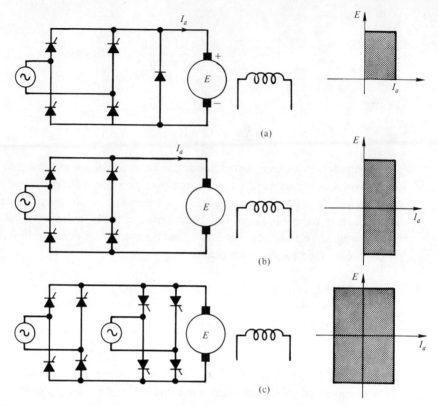

FIGURE 7-14. Single-phase (a) semi- (b) full- and (c) dual converter.

The four types of three-phase converters, and their quadrant operations, are illustrated in Fig. 7.15(a)–(d). In Figs. 7.14(a) and 7.15(b) we have included *freewheeling diodes*, which provide for the flow of armature current when the SCRs are blocking.

Performance of converters is measured in terms of the following parameters:

$$\text{input power factor} = \frac{\text{mean input power}}{\text{rms input volt-amperes}} \qquad (7.9)$$

$$\text{input displacement factor} = \cos \varphi_1 \qquad (7.10)$$

where φ_1 is the angle between the supply voltage and the fundamental component of supply current,

$$\text{harmonic factor} = \frac{\text{rms value of the } n\text{th harmonic current}}{\text{fundamental component of the supply current}} \qquad (7.11)$$

To understand the operation of a converter-fed dc motor, we consider the simplest converter—the single-phase half-wave converter. Figure 7.16(a) shows

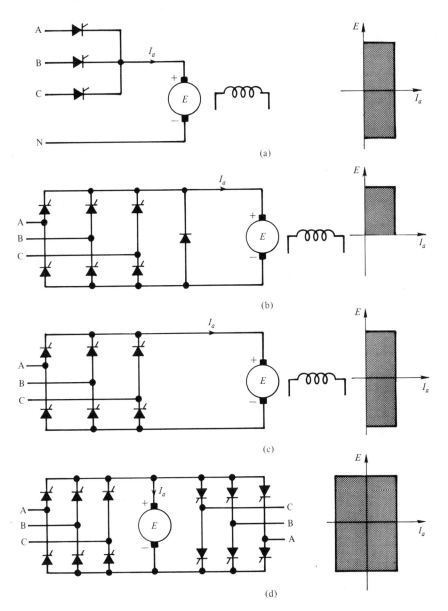

FIGURE 7-15. Three-phase (a) half-wave, (b) semi-, (c) full-, and (d) dual converters.

such a system. The waveforms of motor speed, current, and voltage are given in Fig. 7.16(b). By controlling the firing angle of the SCR, the voltage across the motor armature, and hence its speed, is controlled. In Fig. 7.16(b), α is the firing angle of the SCR. Notice, however, that the armature current does not begin to flow immediately after the SCR is turned on. Only when the line voltage, v_t, becomes greater than the motor voltage, v_m, the armature current

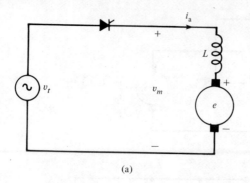

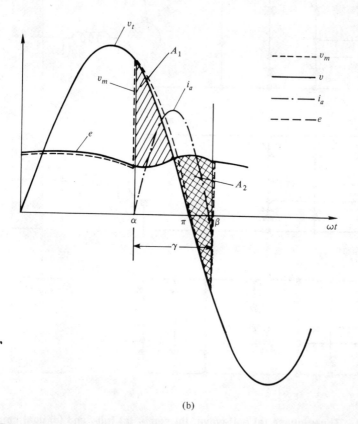

FIGURE 7-16. (a) Half-wave SCR drive; (b) v_m, v_t, i_a, and e waveforms.

begins to flow. The current continues to flow for the period γ, shown in Fig. 7.16(b). This period is also known as the conduction angle, and is determined by the equality of the shaded areas A_1 and A_2 in Fig. 7.16(b). The area A_1 corresponds to the energy stored in the inductance, L, of the motor circuit while the armature current is building up. This stored energy is returned to source during the period the armature current decreases and ultimately becomes zero. The SCR then blocks until it is turned on again. The motor speed is controlled by the firing angle, α. In Fig. 7.16(b), v_t is the waveform of the applied voltage and e is the motor back emf.

A converter-driven motor is analyzed on the basis of averaging over the period of the line voltage, in which case (7.1) to (7.3) are applicable.

EXAMPLE 7.3

A 1-hp 240-V dc motor is designed to run at 500 rpm when supplied from a dc source. The motor armature resistance is 7.56 Ω. The torque constant, k, is 4.23 N-m/A and the back emf constant is 4.23 V/rad/s. This motor is driven by a half-wave converter at 200 V, 50 Hz ac, and draws 2.0 A of average current at a certain load. For the period during which the SCR conducts, the average motor voltage, V_m, is 120 V. Determine (a) the torque developed by the motor; (b) N, the motor speed, in rpm; and (c) the supply power factor.

SOLUTION
(a) Torque $= kI_a = 4.23 \times 2 = 8.46$ N-m
(b) Back emf, $E = 120 - 2.0 \times 7.56 = 104.88$ V
But

$$E = k\Omega_m$$

or

$$104.88 = 4.23\Omega_m = \frac{4.23 \times 2\pi N}{60}$$

Hence

$$N = \frac{60 \times 104.88}{2\pi \times 4.23} = 237 \text{ rpm}$$

(c) Supply volt-amperes $= 200 \times 2 = 400$ VA
Power taken by the motor $= V_m I_a = 120 \times 2$ W
Power factor $= \dfrac{240}{400} = 0.6$

7.4

CONTROLLERS FOR AC MOTORS

A list of the different types of ac motors and the various techniques for speed and torque control of these drives is shown in Fig. 7.17. A control scheme (in

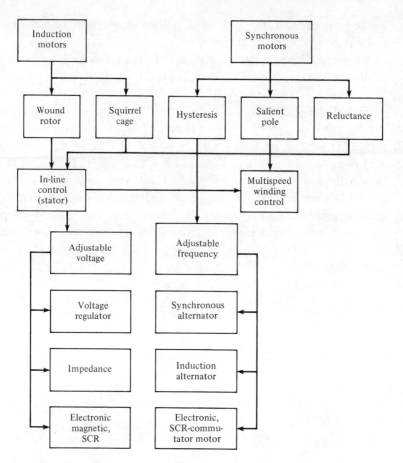

FIGURE 7-17. Ac motors and controls.

the form of a block diagram) for an induction motor is illustrated in Fig. 7.18, in which the power conditioning unit (PCU) includes the energy source (which may even be a dc source), a means of producing an ac variable-frequency source (the inverter) from the available dc source, and some means of controlling the output voltage [the adjustable voltage inverter (AVI), or the pulse-width-modulated (PWM) inverter]. In fact, it would be desirable to keep the voltage-to-frequency ratio (V/f) fixed, as we shall see later.

Inverters. We see from above that the inverter is the backbone of an ac drive system. There is a wide variety of inverter circuits that may be used for a drive motor. Four common types of these inverter types are:

1. Ac transistor inverter.
2. Ac SCR McMurray inverter.

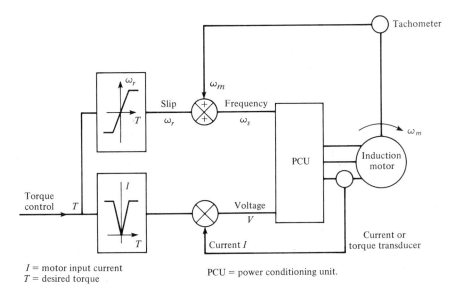

FIGURE 7-18. Block diagram for induction motor control.

3. Ac SCR load-commutated inverter.
4. Ac SCR current inverter (ASCI).

For the present, however, we consider the full-bridge inverter circuit shown in Fig. 7.19, which also shows the voltage and current waveforms. When T1 and T3 are conducting, the battery voltage appears across the load with the polarities shown in Fig. 7.19(b). But when T2 and T4 are conducting, the polarities across the load are reversed. Thus we get a square-wave voltage across the load, and the frequency of this wave can be varied by varying the frequency of gating signals. If the load is not purely resistive, the load current will not reverse instantaneously with the voltage. The antiparallel connected diodes, shown in Fig. 7.19(a), allow for the load current to flow after the voltage reversal.

The principle of the bridge inverter mentioned above can be extended to form the three-phase bridge inverter of Fig. 7.20(a). The gating signals and the output voltages are shown in Fig. 7.20(b). The fundamental components of the line-to-line voltages will form a balanced three-phase system. The antiparallel connected diodes are used to allow flow of currents out of phase with the voltage.

Adjustable Voltage Inverter. In an AVI, the output voltage and frequency can both be varied. The voltage is controlled by including a chopper between the battery and the inverter, whereas the frequency is varied by the frequency of operation of the gating signals. In an AVI, the amplitude of the output decreases with the output frequency and the ratio V/f essentially remains constant over the entire operating range. The AVI output waveform does not

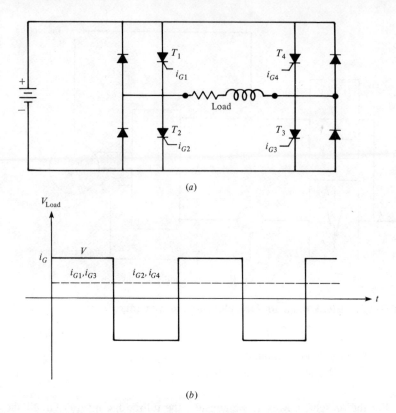

FIGURE 7-19. (a) Single-phase full-bridge inverter; (b) load voltage waveform.

contain second, third, fourth, sixth, eighth, and tenth harmonics. Other harmonic contents as a fraction of the total rms output voltage are:

$$\text{fundamental} = 0.965$$

$$\text{fifth harmonic} = 0.1944$$

$$\text{seventh harmonic} = 0.138$$

$$\text{eleventh harmonic} = 0.087$$

The losses produced by these harmonics tend to heat the motor, and its efficiency decreases by about 5 percent compared to a motor driven by a purely sinusoidal voltage.

Pulse-Width-Modulated (PWM) and Pulse-Frequency-Modulated (PFM) Inverters. The voltage control in PWM and PFM inverters is obtained in a manner similar to that for a chopper, as shown in Fig. 7.10. In a PWM inverter, the output voltage amplitude is fixed and equal to the battery voltage. The voltage is varied by varying the width of the pulse "on time"

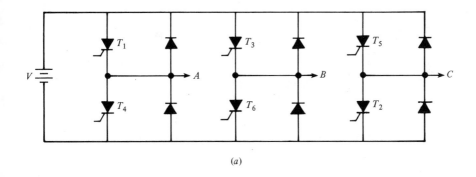

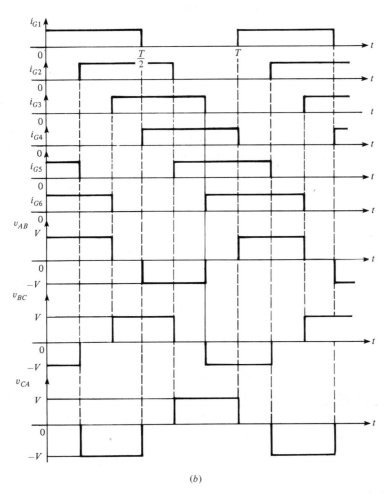

FIGURE 7-20. (a) Three-phase bridge inverter; (b) gating current and output voltage waveforms.

7.4 CONTROLLERS FOR AC MOTORS

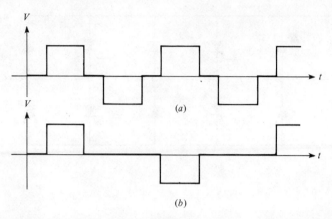

FIGURE 7-21. (a) Full voltage output; (b) half-full voltage output of a single-pulse PWM inverter.

relative to the fundamental half-cycle period, as illustrated in Fig. 7.21. Notice that the full-voltage output wave is similar to that of the AVI output shown in Fig. 7.20(b). However, as the voltage is reduced [Fig. 7.21(b)], the harmonics vary rapidly in magnitude.

Low-frequency harmonics can be reduced by PFM, in which the number and width of pulses within the half-cycle period are varied. PFM waveforms for three different cases are given in Fig. 7.22. Harmonics from the output of a PFM inverter can be reduced by increasing the number of pulses per half-cycle.

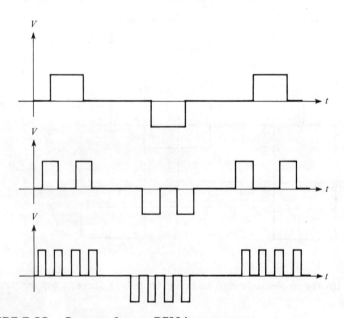

FIGURE 7-22. Outputs from a PFM inverter.

288   **ELECTRIC DRIVES**

But this requires a reduction in the pulse width, which is essentially limited by the thyristor turn-off time and the switching losses of the thyristor. Furthermore, an increase in the number of pulses increases the complexity of the logic system and thereby increases the overall cost of the PFM inverter system.

In comparing an AVI system with a PWM (or PFM) inverter system we observe that whereas the AVI system is efficient but expensive, the PWM inverter is relatively inexpensive but inefficient. Both types of inverter systems are suitable for induction motors. However, for the control of a synchronous motor, a thyristor inverter requires a motor voltage greater than the dc link voltage in order to turn off the thyristors. Also, no diodes are required in the ac portion of the circuit, thus avoiding uncontrolled currents. If a synchronous motor is controlled by a transistor inverter, the motor internal voltage must be less than the battery voltage; otherwise, the diodes in the inverter circuit will conduct, resulting in uncontrollable currents and high losses.

A practical inverter circuit—the McMurray inverter—is shown in Fig. 7.23. This inverter circuit uses auxiliary thyristors to switch a high-Q LC circuit in parallel with the conducting thyristor.

To demonstrate the operation of the McMurray invertor, we redraw the circuit of Fig. 7.23 as a basic single-phase circuit in Fig. 7.24(a). Thyristors T1 and T2 are the main thyristors that carry the load, and T3 and T4 are the auxiliary thyristors. These auxiliary thyristors are used for the commutation of T1 and T2 by switching the high-Q LC pulse circuit in parallel with the conducting thyristor.

Suppose that T1 is initially conducting with the right-hand plate of C_c positive. To turn off T1, T3 is turned on (or gated). The discharge current, i_c, flows

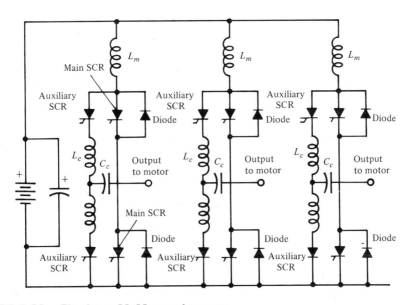

FIGURE 7-23. Thyristor McMurray inverter.

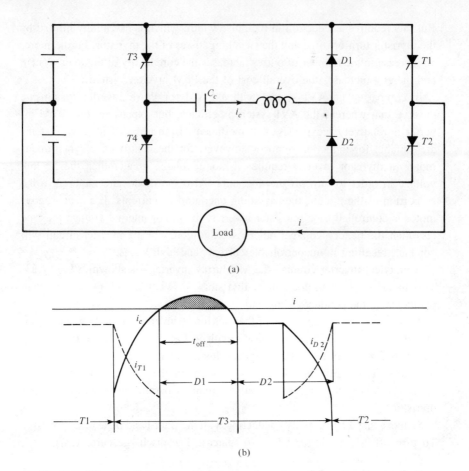

FIGURE 7-24. (a) A 1-phase McMurray inverter; (b) current waveforms.

through T3, C, and L and opposes the load current, i, flowing through T1. Thus T1 is turned off when $i_c = i$. As i_c builds up to exceed i, the excess current flows through the feedback diode D1 and supplies a reverse-bias voltage across T1. The current i_c peaks when the voltage of C_c goes to zero, and then decreases as C_c begins to recharge with reversed polarity. When i_c becomes less than i, the current through D1 stops and the reverse bias across T1 is removed.

After T1 has been turned off, the load current, i, is transferred to D2. When T1 has achieved its forward blocking capacity, T2 is turned on. The sequence of events following are similar to that described in the preceding paragraph. The various current waveforms are illustrated in Fig. 7.24(b).

Figure 7.25 shows a commercially available inverter.

Variable-Frequency Operation of Induction Motors. With a variable-frequency controller, the torque–speed characteristics of an induction motor for several frequencies are shown in Fig. 7.26(a). The dashed-line envelope

FIGURE 7-25. New transistorized AC inverter (460 volts; 5 to 20 horsepower). (Courtesy of Reliance Electric Company)

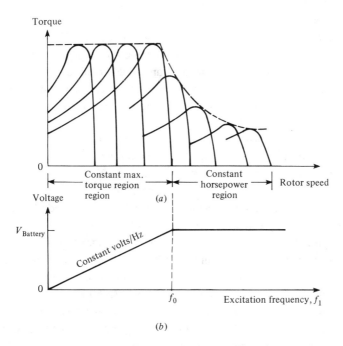

FIGURE 7-26. Typical torque, speed, voltage characteristic for variable-frequency operation of induction motor.

7.4 CONTROLLERS FOR AC MOTORS   **291**

defines two distinct operating regions of constant maximum torque and constant horsepower. In the constant-maximum-torque region, the ratio of applied motor voltage to supply frequency (volts per hertz), is held constant by increasing motor voltage directly with frequency, as shown in Fig. 7.26(b).

The volts/hertz ratio defines the air-gap magnetic flux in the motor and can be held constant for constant torque within the excitation frequency range, and corresponding rotor speeds, extending to frequency f_0. Beyond frequency f_0, the motor voltage cannot be increased to maintain a constant volts/hertz ratio due to the limitation of a finite dc voltage. For motor speeds in the range where f_1 is greater than f_0, the supply voltage is held constant at the battery voltage, and the supply frequency is increased to provide the motor-speed demand. This is the constant maximum horsepower region of operation, since the maximum developed torque decreases nonlinearly with speed.

The envelope illustrated in Fig. 7.26(a) characterizes induction motor operation in a typical electric vehicle drive system. Any speed/torque combination within the envelope can be provided by appropriate voltage and frequency control. For heavy loads, such as vehicle acceleration, the motor is operated in the constant-torque region to provide the torque demanded. For high-speed operation at vehicle cruising speeds, the motor operates in the constant-horsepower region at a frequency to satisfy the demanded speed. Voltage control is usually accomplished in the constant-torque region by pulse-width (duty-cycle) modulation. The motor operates with a fixed-voltage variable-frequency square wave in the constant-horsepower region. Voltage and frequency control are used in both driving and braking operating modes of the vehicle.

The inverter output voltage and current waveforms are rich in harmonics. These harmonics have detrimental effects on the motor performance. Among the most important effects are the production of additional losses and harmonic torques. The additional losses that may occur in a cage induction motor as a result of the harmonics in the input current are summarized as follows:

1. *Primary I^2R losses:* The harmonic currents contribute to the total rms input current. Skin effect in the primary conductors may be neglected in small wire-wound machines, but it should be taken into account in motor analysis when the primary-conductor depth is appreciable.
2. *Secondary I^2R losses:* When calculating the additional secondary I^2R losses, skin effect must be taken into account for all sizes of motor.
3. *Core losses due to harmonic main fluxes:* These core losses occur at high frequencies, but the fluxes are highly damped by induced secondary currents.
4. *Losses due to skew-leakage fluxes:* These losses occur if there is relative skew between the rotor and stator conductors. At 60 Hz the loss is usually small, but it may be appreciable at harmonic frequencies. Since the time-harmonic mmf's rotate relative to both primary and secondary, skew-leakage losses are produced in both members.
5. *Losses due to end-leakage fluxes:* As in the case of skew-leakage losses,

these losses occur in the end regions of both the primary and the secondary and are a function of harmonic frequency.

6. *Space-harmonic mmf losses excited by time-harmonic currents:* These correspond to the losses that, in the case of the fundamental current component, are termed high-frequency stray-load losses.

In addition to these losses, and harmonic torques, the harmonics act as sources of magnetic noise in the motor.

7.5 CHOICE OF MOTOR-DRIVE SYSTEM

The choice of a motor-drive system depends on its ability to do the job efficiently over the projected life at a minimum cost. Whereas one system may be superior to the other in certain respects, the factors governing the choice of a controller are as follows:

1. Power supply available: dc or ac, single-phase or three-phase.
2. Need for inversion.
3. Provision for reversal of rotation of the motor.
4. Overall system cost.
5. Rectifier power factor and efficiency and harmonic generation by the rectifier.

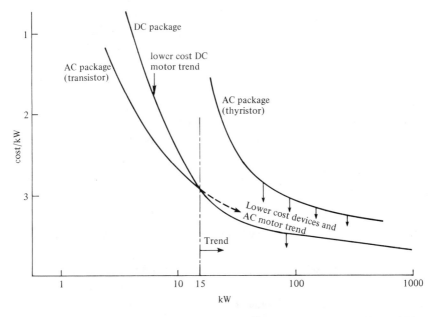

FIGURE 7-27. Cost comparison between ac and dc variable-speed drives. *(Electronics and Power,* Feb. 1983, pp. 135–139, Courtesy IEE, London)

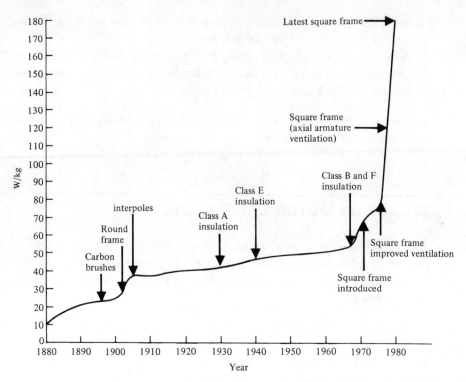

FIGURE 7-28. Progress in output/weight ratio of dc motor over the last 100 years. (*Electronics and Power*, Feb. 1983, pp. 135-139 Courtesy IEE, London)

From a standpoint of initial costs, an ac induction motor of a certain rating is usually less expensive than a dc motor of the same rating. But an inverter to control the ac motor is much more expensive than a dc motor controller. At the present, up to a rating of about 10 kW, ac drives are less expensive than a dc drive. Figure 7.27 shows a cost comparison between ac and dc variable-speed drives. Notice that the trend is for the cost per kilowatt of all drive packages to fall, but at different rates. A breakthrough in the progress of the dc motor output per unit weight, as depicted in Fig. 7.28, has a definite impact on the choice of the motor-drive system.

QUESTIONS

7.1   What are some of the power semiconductors commonly used in electric drives?

7.2   What is an SCR? How does it differ from a diode?

7.3   What is meant by commutation of an SCR?

7.4   What is the purpose of heat sink in power semiconductor circuits?

7.5 How can the speed of a dc motor be controlled by means of power semiconductors if (a) a dc source is available? (b) an ac source is available?

7.6 Explain the function of a free-wheeling diode in a chopper circuit used for dc motor control.

7.7 Define the terms (a) duty cycle, (b) frequency modulation, and (c) pulse-width modulation as used in connection with chopper circuits.

7.8 What is indicated by the quadrant operation of a converter?

7.9 What is the difference between a dual and a full converter?

7.10 A converter is fed from an ac source. Is it used for a dc motor or an ac motor?

7.11 An inverter is supplied from a dc source. Is it used for a dc motor or an ac motor?

7.12 In an adjustable voltage inverter, can the output voltage and frequency both be varied?

7.13 What is the essential difference between a pulse-width-modulated and a pulse-frequency-modulated inverter?

7.14 Can (a) a chopper, (b) a converter, or (c) an inverter be used to control the speed of a permanent-magnet synchronous motor?

7.15 In an inverter-driven induction motor, for a constant torque, it is necessary to keep the volts/hertz ratio constant. Why?

7.16 What are some of the adverse effects of the harmonics in current waveforms driving an induction motor?

7.17 By consulting the latest data sheets from power semiconductor manufacturers, list the maximum voltage and current ratings of (a) a power transistor, (b) a diode, and (c) an SCR.

PROBLEMS

7.1 A dc motor is connected to a 96-V battery through a chopper. If the duty cycle is 45 per center, what is the average voltage applied to the motor?

7.2 A chopper is on for 20 ms and off for the next 50 s. Determine the duty cycle.

7.3 A pulse-width-modulated output waveform from a chopper is shown in Fig. 7P.3. This chopper is connected to a dc motor having an armature resistance of 0.25 Ω and running at 350 rpm. If the motor back emf constant is 0.10 V/rpm, calculate the average armature current.

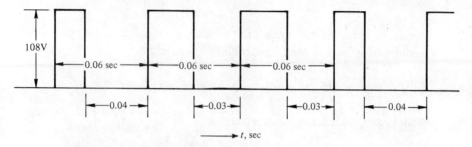

FIGURE 7P.3   PROBLEM 7.3

7.4 Repeat Problem 7.3 if the motor is fed from a frequency-modulated voltage waveform shown in Fig. 7P.4.

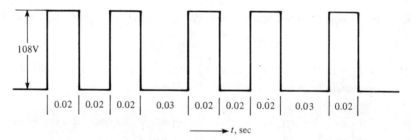

FIGURE 7P.4   PROBLEM 7.4

7.5 A chopper having the following data drives a dc motor: supply voltage 400 V, duty cycle 50 percent, and armature circuit inductance 20 mH. If the chopper frequency is 270 pulses per second, what is the change in the armature current?

7.6 For the motor of Problem 7.5, if the allowable change in the armature current is 5 A, determine the chopper frequency.

7.7 If the chopper of Problem 7.5 is of the form of Fig. 7.12 and can carry 40 A of current, what are the values of the parameters C, L_1, and L_2? Assume that $L_1 = L_2$.

7.8 A 220-V dc shunt motor runs at 750 rpm when supplied from a dc source. When supplied by a half-wave converter at 110 V, 60 Hz ac, the motor draws 1.8 A of average armature current. During the conduction period, the average motor voltage is 60 V. The motor armature resistance is 6.4 Ω and the electromechanical energy conversion constant is 3.5 V/rad/s (or N-m/A). Calculate the motor torque and speed.

7.9 A six-pole 480-V 60-Hz induction motor is driven by an inverter. At 4 percent slip the motor develops 320 Nm torque at rated voltage and frequency. Determine the motor speed (at 4 percent slip) if the motor voltage is reduced to 240 V and the frequency to 30 Hz. Also, estimate the developed torque.

CHAPTER 8

Practical Considerations in Electric Machines

8.1 GENERAL REMARKS

In the preceding chapters we have discussed the theory and performance of various dc and ac electric machines and transformers. We have also considered some of the applications of these machines, and presented a discussion of control of dc and ac motors in Chapter 7. Now that we are familiar with the workings of various electric machines, we wish to consider certain practical aspects of electric machines. We will begin this chapter with nameplate ratings of electric machines and end with motor selection.

8.2 NAMEPLATE RATINGS

Each commercial rotating machine has a nameplate attached to the machine in some manner. Besides the legal and warranty considerations related to the numerical values listed on the nameplate and the usefulness of these values in applying a machine to a specific load, the nameplate is literally the "name" of the machine. It describes the type of machine; its power capability; its speed, voltage, and current characteristics; and some of its environmental limitations. These nameplate parameters give the set of parameters most often used analytically or experimentally to determine machine performance and are widely used in college laboratory experiments and in machine applications. Therefore, a brief description of the meaning of machine nameplate parameters follows.

1. *Power:* This is expressed in terms of the unit horsepower in motors and watts or kilowatts in generators. It refers to the *continuous output* power that the machine can deliver. In machines with variable loading, continuous power is equivalent to average load cycle power, as described in Example 8.1. The continuous power rating is primarily a function of the thermal capacity of the machine, which, in turn, is dependent on the frame configuration, sometimes called the machine "package." There are two basic types of packages in commercial machines, with a number of variations within each type determined by the type of environment and external heat-transfer equipment in which the machine will be operated: *open* (drip-proof, splashproof, externally ventilated, etc.), and *totally enclosed* (nonventilated, fan-cooled, dustproof, water-cooled, encapsulated, etc.).

2. *Speed:* This is expressed in the unit revolutions per minute (rpm). The speed listed on the nameplate depends on the general type of machine: synchronous—synchronous speed; induction—speed at rated power (synchronous minus slip speed at rated power); dc—usually base speed, the maximum speed at which rated torque can be supplied; universal—usually no-load or light-load speed; dc control—usually no-load speed. Many commercial motors are designed to operate from two different voltage source levels or at two different synchronous speeds, and there will be two sets of speed ratings in such cases.

3. *Voltage:* The *nominal* voltage rating of the source to which the machine windings are to be connected; units are volts. In polyphase machines the rated voltage is always expressed as a *line-to-line* voltage. The rated voltage also gives the *insulation level* at which the machine has been constructed and tested. Insulation levels are standardized in commercial machines.

4. *Current:* The *steady-state current* in the armature or power circuit at rated power output and rated speed, expressed in root mean square in ac machines and average amperes in dc commutator machines. In polyphase machines this current is always a *line current*. In dc commutator machines the *field circuit* current rating is the current "full field," that is, the field required for maximum torque at base speed.

5. *Volt-amperes:* In ac motors the input volt-ampere rating is derived from the current and voltage ratings. In single-phase motors, this is

$$\text{rated volt-amperes} = (\text{rated volts})(\text{rated amperes})$$

In three-phase motors,

$$\text{rated volt-amperes} = \sqrt{3}\,(\text{rated volts})(\text{rated amperes})$$

In most ac generators the output volt-ampere rating is stamped on the nameplate in place of the current rating, from which the ouput power factor can be derived. Both the current and volt-ampere ratings of machines are determined primarily by the thermal characteristics of the windings.

6. *Temperature rise:* The maximum safe temperature rise (in degrees Celsius) in the "hot spot" of the machine, which is usually the armature or power windings.
7. *Service factor:* A number that indicates how much over the nameplate power rating the machine can be continuously operated without overheating. It has a value of 1.15 for many commercial motors. In addition, there is a series of *short-time overload ratings* for most commercial motors, which can be obtained from the manufacturer.
8. *Frequency:* The frequency of the supply voltage, in hertz.
9. *Efficiency index:* A recent addition to nameplates of some manufacturers, which gives an indication of minimum and nominal efficiency.
10. *Torque:* Torque generally is listed on nameplates only for control and torque motors. Units vary considerably, although ounce-inches are most common. Rated steady-state torque in power motors can be derived from nameplate power and speed. Two important torque parameters in ac machines—pullout and starting torques at rated voltage—can be obtained through the manufacturer or by means of relatively simple laboratory tests.
11. *Inertia:* This parameter, the inertia of the rotating member, is a nameplate item only for control motors.

```
┌─────────────────────────────────────────────────────────┐
│              GENERAL ELECTRIC                           │
│  ® KINAMATIC        DIRECT CURRENT GENERATOR            │
│  KW  4 1/2      │   RPM  1750   │   VOLTS  125          │
│  ARM AMPS  36                   │   WOUND  COMP.        │
│  FLD AMPS  2.0 AS SHUNT         │ FLD OHMS 25° 47.2 GEN │
│  INSUL CLASS B │ DUTY  CONT.    │ MAX AMBIENT 40°C      │
│     SUIT. AS A 5 HP MTR.  120 V  1800/3600 RPM          │
│                                                         │
│  TYPE  CD256A      │ ENCL  DP   │ INSTR  GEH 2304       │
│  MOD  5CD256627    │            │ SER  FE-1-539         │
│                       ERIE, PA                          │
│  NP 36A424849                          MADE IN U.S.A.   │
└─────────────────────────────────────────────────────────┘
```

FIGURE 8-1. A sample nameplate of a dc machine.

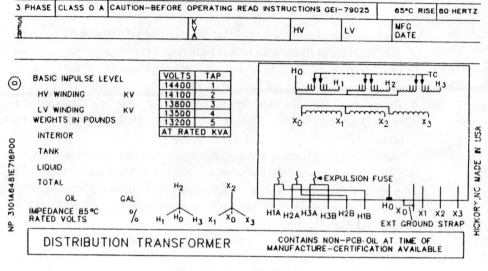

FIGURE 8-2. A sample nameplate of a distribution transformer.

12. *Manufacturer:* The manufacturer's name also appears on the nameplate.
13. *Frame size:* The frame size is the manufacturer's indication of the physical size of the motor and its mounting dimensions. For motors of ratings greater than 1 hp, a standard system of frame numbering and standardized dimensions of these frame sizes have been developed by the National Electrical Manufacturers Association (NEMA).

A typical nameplate of a dc motor is shown in Fig. 8.1. Some of the other information given on some nameplates includes serial number, style, whether it is open or totally enclosed, and code letter to indicate locked-rotor kVA.

A sample nameplate of a three-phase distribution transformer is shown in Fig. 8.2. Notice that the nameplate contains a wealth of information pertinent to the transformer.

8.3

EFFICIENCY AND DUTY CYCLE

Whereas we have treated the topics of efficiency and losses pertinent to the various machine types and transformers, we now wish to discuss efficiency and losses in a broader and general sense. Efficiency of a device is an important factor to be considered in the application of the device. Efficiency can have different meanings in different types of physical systems. In fact, it can have a fairly general meaning that is used in everyday conversation, which is "how well a specific job is done." In mechanical systems use is made of thermal efficiency and mechanical efficiency, which describe the efficiency of two phases of a given process and also "ideal" efficiencies. In the electrical systems that are discussed in this book, efficiency has the following meaning:

$$\eta = \frac{\text{output power or energy}}{\text{input power or energy}} \qquad (8.1)$$

This can also be expressed in terms of mechanical and electrical losses in either energy or power terms as

$$\eta = \frac{\text{output}}{\text{output + losses}} = \frac{\text{input - losses}}{\text{input}} \qquad (8.2)$$

The SI units of power are watts, abbreviated, W; SI units of energy are joules, J; and wattseconds, W-s, or watthours, W-h.

The energy use or efficiency of an electric machine is becoming increasingly significant and is one of the more important design criteria today. Therefore, it is important to know how to calculate the numerator and denominator of the preceding equations. In electric machines, either the numerator or denominator of these equations is a mechanical power or energy. As seen in earlier chapters, mechanical power of a rotating machine can be expressed as

$$P_m = T_{av} \Omega_{av} \qquad (8.3)$$

where
T_{av} = shaft torque, in newton-meters

Ω_{av} = shaft speed, radians per second

On the electrical side of a machine, power is expressed as

$$P_e = VI \cos \theta \quad \text{(sinusoidal)} \qquad (8.4)$$

or

$$P_e = V_{av} I_{av} \quad \text{(dc or pulse)}$$

where
V = terminal voltage, volts

I = terminal current, amperes

θ = power factor angle

In these equations and throughout this book, root-mean-square (rms) parameters have been designated by uppercase, unsubscripted symbols; time-averaged parameters are designated by uppercase symbols and the subscript "av." Power calculated by these equations is *average power*. It is also common to have instantaneous quantities on the right side of these equations, in which lowercase symbols would be used and the power on the left would be referred to as *instantaneous power*. The use of both average and instantaneous power is common in the analysis of electromechanical systems. Energy W is the time integral of power; that is,

$$W = \int p \, dt \tag{8.5}$$

The SI unit of energy is the joule and 1 joule = 1 watt-second.

Duty Cycle. In many applications electric machines are operated at power levels that change with time instead of being at constant power. When this variation in power level can be described as a periodic function, it is known as a *duty cycle*. Duty-cycle variations are used primarily to describe the variations of loading of a motor, but the concept is equally valid for generators, transformers, and other power devices under such conditions. A duty cycle may be expressed in terms of input or output power, shaft torque, armature current, or other machine parameters that describe the load variation. Examples of applications in which the load characteristics can be so described include automatic washing machines, refrigerator compressors, many types of machine tool processes, and electric vehicles traveling through a prescribed driving cycle.

One simple duty cycle is shown in Fig. 8.3 with a time period; it is typical of many motor loads, such as refrigerator compressors. A generalized symbol is used for the magnitude function in Fig. 8.3, since it may represent a power, torque, current, or other parameter. A more complex duty cycle is shown in

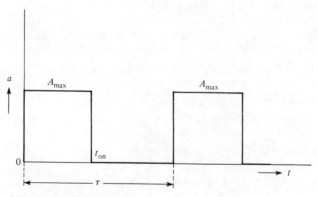

FIGURE 8-3. On/off duty cycle.

Fig. 8.4. This cycle is typical of torque in an electric vehicle, such as a forklift truck, moving through a prescribed set of operations. Note that the function has both positive and negative values. In this application the negative values represent negative or *generator* power, torque, or current and imply a braking operation in which energy is recovered regeneratively. We have also discussed duty cycles in connection with the all-day efficiency of transformers, as in Example 2.8. Dc motor controllers, such as choppers, are also evaluated in terms of duty cycles, as mentioned in Section 7.3.

The principal purposes for the use of duty-cycle analysis are (1) to determine the size or the *rating* of a machine to satisfy a particular load requirement, and (2) to determine *energy efficiency*. The latter purpose is important for machines operating at several power levels, such as described by Fig. 8.4, since the *power efficiency* is generally different at each power level in the duty cycle.

EXAMPLE 8.1

Refer to Fig. 8.3, and let it denote the power absorbed by an electric motor. Let A_{max} correspond to an input power of 900 W during the on-period (t_{on}) of 20 min. The on-period is followed by an off-period of 40 min, and the cycle repeats. Calculate the energy consumed per hour and the average power taken by the motor.

SOLUTION

energy consumed per hour

$$= (900 \times 20 \times 60 + 0 \times 20 \times 60)$$

$$= 108{,}000 \text{ W} - \text{s} \quad (\text{or J})$$

The average power is

$$P_{av} = \frac{108{,}000}{60 \times 60} = 30 \text{ W}$$

■

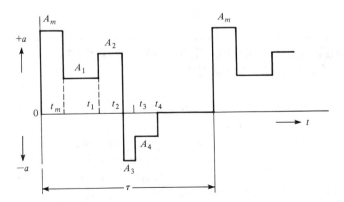

FIGURE 8-4. Duty cycle for electric vehicle with regenerative braking.

8.3 EFFICIENCY AND DUTY CYCLE

EXAMPLE 8.2

In Fig. 8.4, let a be the armature current of the drive motor of an electric vehicle. The currents and their respective durations (in seconds) are:

60 A, $0 \leq t \leq 10$ -40 A, $40 \leq t \leq 45$
25 A, $10 \leq t \leq 30$ -20 A, $45 \leq t \leq 60$
50 A, $30 \leq t \leq 40$ 0 A, $60 \leq t \leq 100$

Determine the average and rms values of this current waveform.

SOLUTION

Time period $t = 100$ s

Algebraic sum of the areas under the positive and negative currents

$= 60 \times 10 + 25 \times 20 + 50 \times 10$
$- 40 \times 5 - 20 \times 15 + 0 \times 40 = 1100$ A s

Average value $= \dfrac{1100}{100} = 11$ A

The rms value is defined as the square root of the mean of the sum of the squares. Hence

$$\text{rms value} = \left[\frac{1}{100}(60^2 \times 10 + 25^2 \times 20 + 50^2 \times 10 + 40^2 \times 5 + 20^2 \times 15)\right]^{1/2}$$

$= 29.58$ A ∎

8.4 RATINGS AND NEMA CLASSIFICATIONS OF ELECTRIC MACHINES

Ratings, as implied here, are closely related to the nameplate ratings discussed in Section 8.2. However, here we restrict ourselves to the ratings that specify the operating limitations on the machine. As in all physical systems, there are physical limitations on electromechanical devices that set bounds on the operation of these devices. An obvious limitation in an electric machine is some maximum speed, operation beyond which will result in damaged bearings, disintegration of the rotating member, or some other such catastrophe. Saturation, the limit on the performance of the magnetic members of the device, is another type of bound; exceeding saturation seldom results in any physical damage, but it limits device performance in many ways. Two other significant bounds on an electric machine operation that result in physical damage or hazardous operation when exceeded are thermal bounds and commutation bounds.

Obviously, a machine should never be operated anywhere near bounds that may cause damage to the machine or injury to operating personnel. Therefore, another set of bounds, known as *machine ratings,* have been developed to guide the machine owner or operator in the proper and safe operation of the machine. Device ratings are always considerably below parameter values that might result in hazardous operation or machine damage. Originally, ratings were somewhat subjective and based largely on past experience and long periods of testing under load. More recently, ratings have sometimes been determined by analytical models of machines and machine subsections. There is a definite legal connotation to a machine rating, and the machine warranties are often invalid if the machine ratings have been exceeded by the user. However, like most rules, machine ratings are designed to be broken or exceeded under certain conditions, and most manufacturers will supply the user with "overload" ratings. This implies that there is a considerable margin between a machine rating and the safety bounds just discussed. Most rotating machines designed for power applications are rugged, long-lived devices and can stand some abuse from the user or the environment. However, this margin of safety built into most rotating machines should be known before using it and should be used only when absolutely essential. In other words, for normal applications, a machine should be operated only at or below its ratings.

It is not the purpose of this book to give a detailed explanation of the scope and meaning of machine ratings. This complex and ever-changing subject is described in a number of publications available in most libraries and from machine manufacturers and users. The *National Electrical Code* describes the use and application of all types of machines, including safety requirements. The size of commercial machines are designated by a parameter known as *frame size*. Frame-size standards are determined by the National Electrical Manufacturers Association (NEMA) and are described in their literature. NEMA has also established a designation, known as *motor classes,* for the general efficiency and starting-torque characteristics of ac motors. Motor Classes A to D exhibit decreasing running efficiency and increasing starting torque with successive alphabetical letter designations.

NEMA classification of the most common types of three-phase induction motors is as follows:

NEMA design B: normal torques, normal slip, normal locked amperes
NEMA design A: high torques, low slip, high locked amperes
NEMA design C: high torques, normal slip, normal locked amperes
NEMA design D: high locked-rotor torque, high slip.

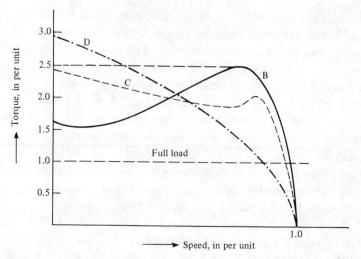

FIGURE 8-5. Torque-speed characteristics of NEMA design B, C and D motors.

Figure 8.5 shows typical NEMA designs B, C, and D motor torque–speed characteristics

The wire used in electric machines is known as *magnet wire*. Wire sizes are most commonly described in terms of numerical designations known as the American Wire Gauge (AWG). Electrical insulation for magnet wire is designated by another alphabetical listing of A to H. Insulation class refers primarily to the safe operating temperature of the insulation and, to a lesser degree, to the structural characteristics and physical strength of the insulation; the temperature range of this designation is from 90°C (Class A) and 110°C (Class B) to 180°C (Class H).

8.5

ECONOMIC CONSIDERATIONS

Economics has always been an important consideration in the use of transformers and electric machines. In the application of machines, the total cost of the machine installation is important; this includes the control system on both the electrical and mechanical side of the machine, the electrical power source, the means of connecting the machine shaft to the load or drive machine, thermal control equipment, and mounting and packaging equipment. The cost and complexity of this auxiliary equipment will often influence the choice of machine to be used in a specific application.

The initial cost of a machine installation must also be balanced by the required energy costs to operate the machine together with maintenance costs. The higher initial cost of an efficient motor and control system can often be compensated by savings in energy costs within a short period compared with the use of a less efficient system. This type of economic trade-off will be more significant in the present era of increasing energy costs. Another factor entering the trade-offs to be considered in choosing a machine for a specific application involves the environment restrictions, which are becoming of increasing importance. For machine installations, this is primarily a matter of audible and electromagnetic noise in both the rotating machine and its control system.

As an example of a typical initial cost versus machine efficiency trade-off, let us consider the single-phase induction motor discussed in Chapter 6. This is one of the most common motor types; it is used in most household appliance applications and is manufactured in huge volumes in a very competitive market. One major use of such motors is in household air-conditioning units. In past years such motors have been designed to compete on the basis of initial cost because of the relatively low cost of electrical energy, and the competitiveness of the marketplace. Consequently, the average full-load efficiency of air-conditioning motors was about 75 percent in 1976. More recently, there has been pressure on motor manufacturers of all types of motors to increase average (or energy) efficiency of electric machines, since a surprisingly significant percentage of the energy used in an industrialized nation such as the United States is wasted in the losses of electric machines. For a motor operating at a set power level and duty cycle, such as the household air-conditioning motor, there are relatively few choices available for increasing motor efficiency. The most obvious approach is to reduce the motor losses, which requires a redesign of the motor. Magnetic losses can be reduced by using improved magnetic materials—either higher-quality steel or thinner laminations, both of which add cost to a motor. Ohmic loss can be reduced by the use of conductors of larger cross section or, in some situations, by redesigning the winding layout. About the only component of mechanical loss that can be reduced is the bearing loss, through the use of less lossy bearings which, similarly, is achieved with a cost penalty. In some cases a total redesign of the motor may result in improved efficiency but when one considers that this class of motors is generally based on optimal designs for given frame sizes and applications, there is probably little efficiency improvement that could be picked up in this manner. Therefore, efficiency of appliance motors can be improved almost solely by increasing the initial cost of the motor (improved materials, new lamination designs, reduced air gap, better bearings, etc.). The next example illustrates such a trade-off.

EXAMPLE 8.3

A major manufacturer of home and office air conditioners uses a $\frac{1}{2}$-hp single-phase induction motor that has an efficiency of 72 percent at its average power output level. Several large-volume customers indicate that for patriotic reasons and because of government subsidy of energy-saving installations, they would be willing to pay a larger initial investment in air conditioners if they could recover this increased investment during the warranty period, which happens to be 2 years. The typical office air conditioner in which they are interested runs approximately 8 hours per day during 140 equivalent days of an Atlanta year. The motor supplied to the manufacturer has a wholesale or OEM cost of $45. If the motor supplier could achieve an average efficiency of 85 percent by improving materials and design, how much cost differential could be added to the wholesale motor cost and still satisfy the customer's request?

SOLUTION

The energy used by one present motor is $(373/0.72)(140 \times 8 \times 2)/1000 = 1160$ kWh in 2 years. At an efficiency of 85 percent, this would be reduced to $(373/0.85)(140 \times 8 \times 2) = 983$ kWh in 2 years. The energy savings (Atlanta had an average electrical energy cost of 5.6 cents/kWh in 1979) would be $(1160 - 093)(0.056) = \$9.91$. This could be added to the initial cost of motor manufacture. ∎

Motor efficiency can be improved in other ways than those just discussed by changing the motor control or other conditions *external* to the motor. In some cases changing the nominal voltage rating may improve motor efficiency but, of course, the use of such an optimal voltage may not be possible because of unavailability of the required voltage source. In general, voltage optimization is much more feasible in dc commutator motor applications than for ac motors because of the flexibility of dc voltage sources. However, ac motors are amenable to many other techniques that alter the equivalent voltage source. These techniques are based on increasing the average power factor at which the motor operates. Note that motor power factor improvement results in secondary efficiency improvements in the overall power system supplying the motor due to reduced line currents in transmission and distribution lines and in the source generators. Recently, a simple electronic controller for single-phase induction motors that increases average motor power factor has been developed by the NASA Marshall Laboratory and is available commercially.

8.6

THERMAL CONSIDERATIONS

The losses—electrical, magnetic, and mechanical—generate heat in electric machines. Consequently, the temperature surrounding the origin of the loss rises. The overall rise of temperature depends on the losses, that is, the rate of heat generation, the rate of heat transfer, and the thermal capacity of the machine. The machine rating is limited by its temperature rise above the ambient temperature (as discussed in Section 8.4). The temperature rise in a machine is ideally given by

$$T = T_m(1 - e^{-t/\tau}) \tag{8.6}$$

where T_m is the final temperature rise and τ is the thermal time constant. The final temperature and the thermal time constant depend on the following factors:

- p = rate of heat development, which is also the power dissipated as heat (or losses), J/s or W
- G = weight of active parts of machine, kg
- h = specific heat, J/kg/°C
- S = cooling surface area, m²
- λ = specific heat dissipation or emissivity, J/s/m² of surface per °C between surface and ambient cooling medium

In particular, it has been found that

$$T_m = \frac{\text{power dissipated as heat or losses}}{(\text{emissivity})(\text{surface area of power dissipation})} = \frac{p}{S\lambda} \quad °C$$

$$\tau = \frac{(\text{thermal capacity})(\text{cass of the body generating heat})}{(\text{emissivity})(\text{surface area of the body})} = \frac{Gh}{S\lambda} \quad s$$

Thermal time constants vary from a few seconds to a few hours.
The cooling, or temperature drop, is given by

$$T = T_m e^{-t/\tau} \tag{8.7}$$

where T_m is the initial temperature when cooling began.

EXAMPLE 8.4

A transformer has a loss of 5 kW while operating on full-load. The transformer weighs 800 kg, has a thermal capacity of 500 J/kg/°C, and has a surface area of 5 m². The emissivity is 10 W/m²/°C. Determine the variation of temperature with time.

SOLUTION

The temperature–time relationship is given by (8.6), where

$$T_m = \frac{5000}{10 \times 5} = 100°C$$

$$\tau = \frac{500 \times 800}{10 \times 5} = 8000 \text{ s} = 2.22 \text{ h}$$

Hence, from (8.6),

$$T = 100(1 - e^{-t/2.22}) \quad °C$$

where t is in hours.

8.6 THERMAL CONSIDERATIONS

EXAMPLE 8.5

The temperature rise of an electric machine is 15°C after 1 h and 24°C after 2 h of continuous operation. Determine, approximately, the final steady temperature rise.

SOLUTION

Let τ be the thermal time constant. Then, from (8.6), we get

$$15 = T_m(1 - e^{-1/\tau})$$

$$24 = T_m(1 - e^{-2/\tau})$$

These equations yield

$$\frac{15}{24} = \frac{1 - e^{-1/\tau}}{1 - e^{-2/\tau}}$$

from which $\tau = 1.96$ h. Consequently,

$$T_m = \frac{15}{1 - e^{-1/1.96}} = 37.5°C \qquad \blacksquare$$

To limit the temperature rise, electric machines are provided with some means of cooling. The cooling may be by conduction, convection, or radiation, and the machine may have either natural cooling or forced cooling. Many self-ventilated machines have fans mounted on the rotor. External fans are used in forced-ventilated machines. Totally enclosed machines may or may not have separate fans. A cooling-air circuit showing radial and axial flow of air is given in Fig. 8.6. The fan mounted on the rotor draws air through radial ducts. Smaller machines do not require radial ducts, and the flow of air, as drawn by the fan, is only axial.

Whereas a detailed thermal analysis of an electric machine is beyond the scope of this book, the following relationships are useful in approximately estimating the maximum temperature rise, T_m, of a given surface.

$$T_m = \frac{\text{cooling coefficient}}{\text{area of dissipated surface}} \times \text{power dissipated}$$

For cylindrical rotors the cooling coefficient takes a value of [0.015 to 0.035/(1 + 0.1u)], where u is the armature peripheral speed, and smaller values are for large open machines. It has been found that in a totally enclosed machine the outer-surface dissipation is of the order of 12 W/m²/°C.

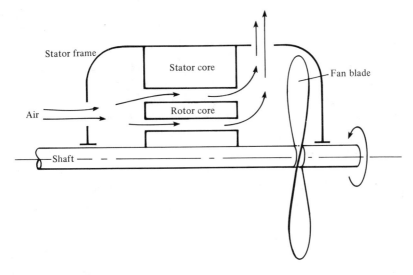

FIGURE 8-6. Cooling-air flow in an electric machine.

8.7

MOTOR SELECTION

Numerous factors enter into the choice of a motor to drive a load. Some of these factors are:

1. *Supply available:* whether dc or ac; single-phase or three-phase; voltage and current ratings of the supply.
2. *Environment:* nonhostile or hazardous, such as oil, corrosive chemicals, dust, and water vapor, and ambient temperature.
3. *Mounting:* vertical or horizontal.
4. *Size and shape:* if the motor can be accommodated in the available space.
5. *Torque:* speed characteristic of the load.
6. *Speed:* speed range and ease of control.
7. *Duty cycle:* frequency of starts and stops in a given time; continuous, intermittent, or varying duty.
8. *Acceleration and deceleration characteristics.*
9. *Economic considerations:* initial cost and operating expenses of the motor (and controller, if required).
10. *Overload considerations:* continuous-duty general-purpose motors can take momentary overloads not exceeding 75 percent of the breakdown torque.
11. *Availability of control equipment.*

Clearly, with so many factors influencing the choice of a drive, it is almost impossible to uniquely select a motor/controller system for a specific job. Nevertheless, we list below a number of applications of dc and ac motors to show the trend.

Steel Rolling Mills. The torque and speed requirements in processing a billet in a steel rolling mill are quite stringent. The drive motor has to undergo severe and rapid torque and speed fluctuations. For such applications SCR-controlled dc shunt motors have been found to be most suitable.

Electric Traction. For electric locomotives, dc series motors are best suited. Phase-controlled SCR power-conditioning equipment is often used to control the motor torque and speed. Recently, dc shunt motors and three-phase induction motors have also been used in electric traction.

Paper Mills. In papermaking machines, accurate speed control of drive motors is essential. For instance, paper speeds in the range 0.1 to 10 m/s, with a steady-state speed variation of ±0.05 percent, may be required. For such applications, SCR-controlled dc shunt motors are appropriate.

Electric Vehicles. Electric vehicles, such as electric cars, have on-board batteries as power sources. Thus chopper-controlled dc series or shunt motors are commonly used. However, three-phase induction and synchronous motors, in conjunction with inverters, are also being considered suitable for electric vehicles.

Pumps, Blowers, and Compressors. Loads such as water supply pumps, compressors, and blowers require a starting torque of 100 to 150 percent of full-load torque, 200 to 250 percent of full-load torque during short time overloads, constant speed, and good speed regulation. For such applications, NEMA design B induction motors are commonly used. Variable-frequency variable-voltage inverters are used for the control of such motors.

Reciprocating Pumps, Crushers, and Ball Mills. Loads such as those in ball mills and crushers require 200 to 300 percent torque for starting, because of high inertia, standstill friction, back pressure, and so on. But for running, no more than full-load torque is necessary. NEMA design C induction motors are quite adequate for such purposes.

Punch Presses, Cranes, Hoists, and Shears. Such loads as punch presses, cranes, and shears are classified as intermittent loads. They require frequent start, stop, and reverse operations. The starting torque needed is about 300 percent of the full-load torque and the maximum running torque is between 200 and 300 percent of the full-load torque. For such applications, high-slip, high-torque NEMA design D motors are often used.

Household Appliances. For appliances such as blenders, mixers, hand tools, and vacuum cleaners, universal motors are commonly used. Shaded-pole motors are used for fan applications, unit heaters, hair driers, and so on. For washing machines, air conditioners, and refrigerators, single-phase, general-purpose (or high-torque) induction motors are most commonly used.

Constant-Speed Applications. Synchronous motors—reluctance, hysteresis, permanent magnet, and field-excited types—are ideally suited for loads to be driven at constant speeds. For small-power applications, such as electric clocks, turntables, and timing devices, reluctance and hysteresis motors are used. Applications of large synchronous motors is dependent on a number of factors in choosing between synchronous and induction motors. From experience, synchronous motors are less expensive than cage-type induction motors for ratings exceeding 1 hp/rpm. In addition, synchronous motors have higher efficiencies and better power factors compared to induction motors. Some of the factors against synchronous motors are:

1. Necessity for excitation source and means of field control.
2. Relative low torque/kVA starting efficiency.
3. Slightly greater maintenance cost.

General rules as a preliminary guide for selecting a synchronous or an induction motor are as follows:

1. Below a speed of 500 rpm, a synchronous motor is better suited than an induction motor.
2. If low-starting kVA and controllable torque are required, a synchronous motor is a better choice.
3. Above a rating of 5000 hp, a synchronous motor should be the first choice.

Figure 8.7 shows general ratings of applications of large synchronous and induction motors.

8.8 VARYING LOAD APPLICATIONS

In the preceding section we listed a range of applications of dc and ac motors. In many applications the load on the motor fluctuates and undergoes repeated cycles. The motor size selection in such cases is based on the concept of rms horsepower (rms hp), which is defined as that equivalent steady-state horsepower

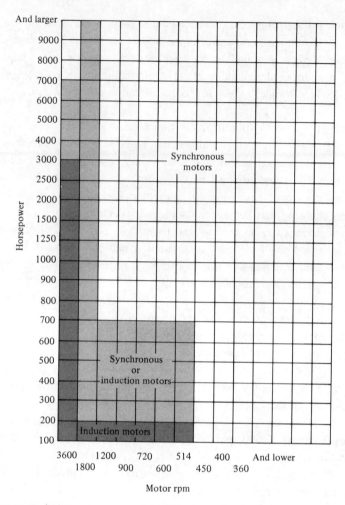

FIGURE 8-7. General areas of application of synchronous motors and induction motors.

which would result in the same temperature rise as that of the varying load cycle. The rms hp is calculated as follows.

EXAMPLE 8.6

Over a cycle period of 1 h, a motor carries a load of 50 hp for 15 min, 30 hp for 25 min, 10 hp for 10 min, and standstill for 10 min. Assuming that the heat dissipation is 100 percent effective while the motor is running, but is reduced to 30 percent at standstill, determine the rms hp and select a suitable induction motor.

SOLUTION

$$\text{rms hp} = \sqrt{\frac{(\text{hp})^2 \times \text{time for 1 cycle}}{\text{effective cooling time}}} \qquad (8.8)$$

$$(\text{hp})^2 \times \text{time for 1 cycle}$$
$$= (50)^2 \times 15 + (30)^2 \times 25$$
$$+ (10)^2 \times 10 + 0 \times 10$$
$$= 61{,}000$$

$$\text{effective cooling time} = 15 + 25 + 10 + 0.3 \times 10$$
$$= 53 \text{ min}$$

$$\text{rms hp} = \sqrt{\frac{61{,}000}{53}} = 33.9 \text{ hp}$$

For thermal limitations, a 40-hp motor should be satisfactory. Considering the peak-load requirement, we have

$$\frac{\text{max. hp}}{\text{rms hp}} = \frac{50}{40} = 1.25 \text{ or } 125 \text{ percent}$$

Hence a NEMA design B 40-hp induction motor is adequate for the job. ■

8.9 INSULATION CONSIDERATIONS

Insulators, such as glass, polymers, and mica, have resistivity in the range 10^{10} to 10^{17} Ω-cm. Insulations used in electric machines are composite, in that the insulation comprises elements that serve as dielectrics (insulators), as well as aid in bonding, reinforcement, sealing, and protection. Thus, for an insulator to be useful, it must have acceptable thermal and mechanical properties. Table 8.1 shows some of the properties of insulating materials used in electric machines.

The quality of coil insulation is evaluated from the following tests:

1. *Voltage breakdown strength:* determined by short-time increases in applied voltage, carried out to failure.
2. *Voltage endurance strength:* determined by extended life tests. The duration of the test at each voltage level is progressively decreased as the voltage is increased.

TABLE 8.1 Properties of Insulating Materials

Material	Breakdown Voltage (V/m)	Resistivity (Ω-cm)	Dielectric Constant at 60 Hz	Power Factor at 60 Hz
Mica flake	1500–3000	10^{17}	7.5	0.0025
Epoxy	400	10^{16}	3.6	0.02
Polyester	250–500	10^{14}	4.1–5.2	0.03
Polyethylene	460	10^{13}	2.3	0.0002
Silicone	185	10^{12}	3.7	0.001

3. *Temperature endurance:* determined by functional tests. Actual coil samples are mounted in slots simulating the core and subjected to high temperatures in a controlled oven with a periodic cycle of humidification, mechanical vibration, and electrical tests.
4. *Moisture resistance:* determined by immersing coils in a water bath and applying a voltage to ground equal to the rated voltage. The completely wound stator may be submerged in a test tank for such tests. In this case the applied voltage is 1.15 times rated value.
5. *Chemical resistance:* determined by testing samples after long-term immer-

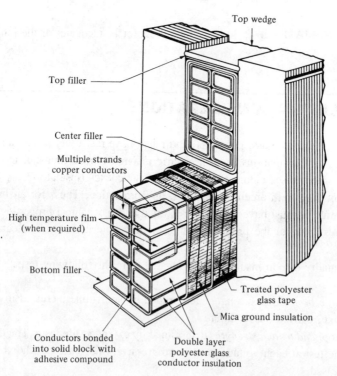

FIGURE 8-8. Cross section of a treated vacuum pressure impregnated coil.

sion in chemical containments to determine the effects of acids, alkalis, oils, and so on.
6. *Abrasion resistance:* determined by testing after subjecting the coil insulation to a stream of abrasive particles blown by compressed air under specific pressure and angle of application.

The stator coils of large machines are wound and formed according to design specifications. The straight portions of coils are insulated with multiple layers of mica-paper wrapper. Coil lead pockets are then filled with an epoxy compound. The ends and leads are also insulated with multiple layers of mica-paper tape. The entire coil is covered with a half overlapped layer of specially treated polyester/glass tape. Coils are then wedged into slots to a snug fit. Coil ends are evenly spaced and supported by placing treated polyester felt pads between coils. The completely wound stator is finally subjected to a process known as vacuum-pressure impregnation resulting in void-free coil insulation. Figure 8.8 shows the details of a coil insulation.

QUESTIONS

8.1 What is the purpose of a nameplate on an electric machine?

8.2 What is meant by energy efficiency of a machine?

8.3 Under what circumstances is it more important to consider the energy efficiency rather than power efficiency for an electric machine application?

8.4 Why is power factor important in determining the efficiency of a machine?

8.5 Does the duty cycle have an effect on the rating of a machine? Explain.

8.6 What are the various designs of induction motors, and their respective characteristics, according to NEMA classification?

8.7 Is a NEMA design B induction motor suitable to start a big load requiring a high starting torque?

8.8 Among Class A, B, and H insulations, which is rated for highest temperature?

8.9 In applying energy-efficient motors, is it important to consider their initial costs from an economic viewpoint? Why?

8.10 What is the significance of the thermal time constant in determining the temperature rise in an electric machine?

8.11 Does the final temperature rise of the rotor of a machine depend on the rotor volume or its surface area?

8.12 What are the causes of temperature rise in an electric machine?

8.13 List the important factors that govern the choice of an electric motor for a given application.

8.14 Name the suitable types of motors for the following applications: electric cranes; wind tunnel drive; large compressors; electric clocks; turntables; and large rotary pumps.

8.15 Define rms horsepower and explain how it is related to the duty cycle of a motor.

8.16 What characteristics are desirable in insulation for a high-voltage machine?

PROBLEMS

8.1 An electric motor runs at 1800 rpm and delivers a torque of 420 N-m. The total losses at 1800 rpm are 9600 W. Calculate the motor efficiency.

8.2 The output torque and speed of a motor over a 40-min period vary in the manner shown in Fig. 8P.2. What is the average power output of the motor?

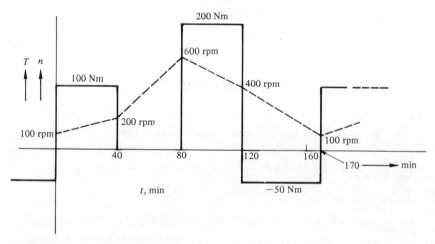

FIGURE 8P.2   PROBLEM 8.2

8.3 The power absorbed by a motor over a 1-h period is shown in Fig. 8P.3. The losses for this period are also given in Fig. 8P.3. Determine (a) the energy efficiency and (b) the average power efficiency.

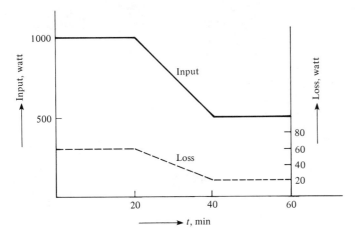

FIGURE 8P.3 PROBLEM 8.3

8.4 A motor runs for 10 h per day during 230 equivalent days per year, while delivering an average power of 6 hp at an average efficiency of 75 percent during this period. The cost of this motor is $465. If this motor is to be replaced by a motor having an average efficiency of 87 percent but costing $535, how long will it take to recover the cost differential of the two motors if electrical energy rate is 6.2 cents/kWh?

8.5 An electric machine weighing 1000 kg has a total loss of 8.4 kW on full-load. The effective cooling surface area of the machine is 6.2 m². The machine has a thermal capacity of 500 J/kg/°C and an emissivity of 9.2 W/m²/°C. Determine the average machine temperature after it has operated for 1 h.

8.6 The temperature rise of an electric machine is 20°C after 1 h and 32°C after 2.2 h. Determine the thermal time constant and the final temperature rise.

8.7 An induction motor carries a load of 100 hp for 10 min, 50 hp for 20 min, and 10 hp for 30 min. During this period of operation (1 h) the cooling is effective 100 percent. Calculate the rms horsepower of the motor. Specify a NEMA design motor suitable for this load.

References

1. **Fitzgerald, A. E., C. Kingsley,** and **A. Kusko,** *Electric Machinery,* 3rd ed., McGraw-Hill, New York, 1971.
2. **Jaeschke, R. L.,** *Controlling Power Transmission Systems,* Penton/IPC, Cleveland, Ohio, 1978.
3. **Kosow, I. L.,** *Electric Machinery and Transformers,* Prentice-Hall, Englewood Cliffs, N.J., 1972.
4. **Langsdorf, A. S.,** *Theory of Alternating-Current Machinery,* 2nd ed., McGraw-Hill, New York, 1955.
5. **Nasar, S. A.,** *Schaum's Outline of Theory and Problems in Electric Machines and Electromechanics,* McGraw-Hill, New York, 1981.
6. **Nasar, S. A.,** and **L. E. Unnewehr,** *Electromechanics and Electric Machines,* 2nd ed., Wiley, New York, 1983.
7. **Say, M. G.,** and **E. O. Taylor,** *Direct Current Machines,* Halsted Press, New York, 1980.
8. **Sen, P. C.,** *Thyristor Dc Drives,* Wiley-Interscience, New York, 1981.
9. **Stein, R.,** and **W. T. Hunt, Jr.,** *Electric Power System Components,* Van Nostrand Reinhold, New York, 1979.
10. **Sugandhi, R. K.,** and **K. K. Sugandhi,** *Thyristors—Theory and Applications,* Halsted Press, New York, 1981.
11. **Unnewehr, L. E.,** and **S. A. Nasar,** *Electric Vehicle Technology,* Wiley-Interscience, New York, 1982.
12. **Veinott, C. G.,** *Fractional Horsepower Electric Motors,* 2nd ed., McGraw-Hill, New York, 1948.

APPENDIX A

Dimensions, Weight, and Resistance of Pure Copper Wire

AWG Number	Diameter (in.)	Area, d^2 (circular mils; 1 mil = 0.001 in.)	Pounds per 1000 ft Bare Wire	Length (ft/lb)	Resistance at 77°F (Ω/1000 ft)
	1.151	1000000.	3090.	0.3235	0.0108
	1.029	800000.	2470.	0.4024	0.0135
	0.963	700000.	2160.	0.4628	0.1054
	0.891	600000.	1850.	0.5400	0.0180
	0.814	500000.	1540.	0.6488	0.0216
	0.726	400000.	1240.	0.8060	0.0270
Stranded	0.574	250000.	772.	1.30	0.0341
0000	0.4600	211600.	640.5	1.56	0.490
000	0.4096	167800.	507.9	1.97	0.0618
00	0.3648	133100.	402.8	2.48	0.0871
0	0.3249	105500.	319.5	3.13	0.0983
1	0.2893	83690.	253.3	3.95	0.1239
2	0.2576	66370.	200.9	4.96	0.1563
3	0.2294	52630.	159.3	6.28	0.1970
4	0.2043	41740.	126.4	7.91	0.2485
5	0.1819	33100.	100.2	9.98	0.3133
6	0.1620	26250.	79.46	12.59	0.3951
7	0.1443	20820.	63.02	15.87	0.4982
8	0.1285	16510.	49.98	20.01	0.6282
9	0.1144	13090.	39.63	25.23	0.7921
10	0.1019	10380.	31.43	31.82	0.9989
11	0.09074	8234.	24.92	40.12	1.260
12	0.08081	6530.	19.77	50.59	1.588
13	0.07196	5178.	15.68	63.80	2.003
14	0.06408	4107.	12.43	80.44	2.525
15	0.05707	3257.	9.86	101.4	3.184
16	0.05082	2583.	7.82	127.9	4.016
17	0.04526	2048.	6.20	161.3	5.064
18	0.04030	1624.	4.92	203.4	6.385
19	0.03589	1288.	3.90	256.5	8.051
20	0.03196	1022.	3.09	323.4	10.15
21	0.02846	810.1	2.45	407.8	12.80
22	0.02535	642.4	1.95	514.2	16.14
23	0.02257	509.5	1.54	648.4	20.36
24	0.02010	404.0	1.22	817.7	25.67
25	0.01790	320.4	0.970	1031.0	32.37
26	0.01594	254.1	0.769	1300.0	40.81
27	0.01420	201.5	0.610	1639.0	51.47
28	0.01264	159.8	0.484	2067.0	64.90
29	0.01126	126.7	0.384	2607.0	81.83
30	0.01003	100.5	0.304	3287.0	103.2
31	0.00893	79.70	0.241	4145.0	130.1
32	0.00795	63.21	0.191	5227.0	164.1
33	0.00708	50.13	0.152	6591.0	206.9
34	0.00631	39.75	0.120	8310.0	260.9
35	0.00562	31.52	0.095	10480.0	329.0
36	0.00500	25.00	0.076	13210.0	414.8

APPENDIX B

Unit Conversion

Symbol	Description	One: (SI Unit)	Is Equal to: (English Unit)	(CGS Unit)
B	Magnetic flux density	tesla (T) ($=1$ Wb/m^2)	6.452×10^4 lines/in.^{2}	10^4 G
H	Magnetic field intensity	ampere per meter (A/m)	0.0254 A/in.	0.004π Oe
φ	Magnetic flux	weber (Wb)	10^8 lines	10^8 Mx
D	Viscous damping coefficient	newton-meter-second (N-m-s)	0.73756 lb-ft-sec	10^7 dyn-cm-sec
F	Force	newton (N)	0.2248 lb	10^5 dyn
J	Inertia	kg-square meter	23.73 lb-ft^2	10^7 g-cm^2
T	Torque	newton-meter (N-m)	0.73756 ft-lb	10^7 dyn-cm
W	Energy	joule (J)	1 W-sec	10^7 ergs

APPENDIX C

Data on DC Machines

Connections for DC Generators

Shunt Generator

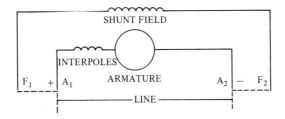

Compound Generator

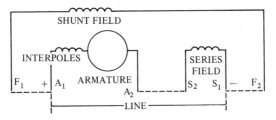

All connections for counterclockwise rotation facing commutator end. For clockwise rotation interchange A_1 and A_2.

For the generators above, the shunt field may be either self-excited or separately excited. When self-excited, connections should be made as shown. When separately excited, the shunt field is isolated from the other windings. When separately excited, the same polarities must be observed for given rotation.

Connections for DC Motors

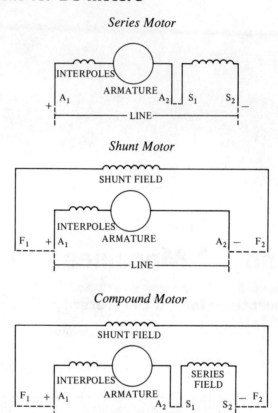

All connections for counterclockwise rotation facing commutator end. For clockwise rotation interchange A_1 and A_2.

When the shunt field is separately excited, the same polarities must be observed for a given rotation.

Approximate Full-Load Terminal Currents for DC Shunt Motors

hp	120 V	240 V	hp	240 V
¼	3.1	1.6	15	55
⅓	4.1	2.0	20	72
½	5.4	2.7	25	89
¾	7.6	3.8	30	106
1	9.5	4.7	40	140
1½	13.2	6.6	50	173
2	17	8.5	60	206
3	25	12.5	75	255
5	40	20	100	341
7½	58	29	125	425
10	76	38	150	506
			200	675
			OVER 200 Approx. A/hp 3.4	

DATA ON DC MACHINES

APPENDIX D

Data on AC Machines

Connections for Nine-Lead, Three-Phase Motors

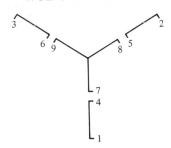
WYE-CONNECTED

Voltage	Line 1	Line 2	Line 3	Together
Low	1 & 7	2 & 8	3 & 9	4&5&6
High	1	2	3	4&7, 5&8, 6&9

DELTA-CONNECTED

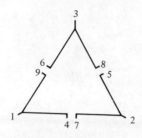

Voltage	Line 1	Line 2	Line 3	Together
Low	1&6&7	2&4&8	3&5&9	None
High	1	2	3	4&7, 5&8, 6&9

Connections for Two-Speed, Three-Phase Motors

CONSTANT TORQUE

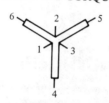

Speed	Line 1	Line 2	Line 3	
High	4	5	6	1-2-3 Together
Low	2	3	1	4-5-6 Open

CONSTANT HORSEPOWER

Speed	Line 1	Line 2	Line 3	
High	4	5	6	1-2-3 Open
Low	2	3	1	4-5-6 Together

VARIABLE TORQUE

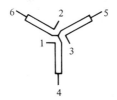

Speed	Line 1	Line 2	Line 3	
High	4	5	6	1-2-3 Together
Low	2	3	1	4-5-6 Open

Motor Full-Load Currents

	Three-Phase AC Induction Type—Squirrel Cage and Wound Rotor							Single Phase	
hp	115 V	200 V	230 V	460 V	575 V	2300 V	4160 V	hp	115 V	230 V
½	4	2.3	2	1	.8			⅙	4.4	2.2
¾	5.6	3.2	2.8	1.4	1.1			¼	5.8	2.9
1	7.2	4.15	3.6	1.8	1.4			⅓	7.2	3.6
1½	10.4	6	5.2	2.6	2.1			½	9.8	4.9
2	13.6	7.8	6.8	3.4	2.7			¾	13.8	6.9
3		11	9.6	4.8	3.9			1	16	8
5		17.5	15.2	7.6	6.1			1½	20	10
7½		25	22	11	9			2	24	12
10		32	28	14	11			3	34	17
15		48	42	21	17			5	56	28
20		62	54	27	22			7½	80	40
25		78	68	34	27			10	100	50
30		92	80	40	32					
40		120	104	52	41					
50		150	130	65	52					
60		177	154	77	62	16	8.9			
75		221	192	96	77	20	11			
100		285	248	124	99	26	14.4			
125		358	312	156	125	31	17			
150		415	360	180	144	37	20.5			
200		550	480	240	192	49	27			
Over 200 hp Approx. A/hp		2.75	2.40	1.20	.96	.24	.133			

Three-Phase Synchronous Motors Unity Power Factor

hp	440 V	550 V	2300 V	4160 V	hp	440 V	550 V	2300 V	4160 V
100	106	85	20	11.2	400	420	336	80.5	44.4
125	132	106	25	14.0	500	525	420	100	55.5
150	158	127	30	16.7	600	630	505	120	66.5
200	210	168	40	22.2	700	735	588	141	77.7
250	262	210	50	27.7	800	840	671	161	88.8
300	315	252	60	33.3	900	945	755	181	100
350	368	295	70.5	38.9	1000	1050	840	192	110

Maximum Locked-Rotor Currents—Three-Phase Motors

hp	Voltage					
	200	220/230	440/460	550/575	2300	4160
$\frac{1}{2}$	23	20	10	8		
$\frac{3}{4}$	29	25	12.5	10		
1	34.5	30	15	12		
$1\frac{1}{2}$	46	40	20	16		
2	57.5	50	25	20		
3	73.5	64	32	25		
5	106	92	46	37		
$7\frac{1}{2}$	146	127	63	51		
10	186	162	81	65		
15	267	232	116	93		
20	334	290	145	116		
25	420	365	182	146	35	19
30	500	435	217	174	41	23
40	667	580	290	232	55	30
50	834	725	362	290	69'	38
60	1000	870	435	348	83	46
75	1250	1085	592	435	104	57
100	1670	1450	725	580	139	76
125	2085	1815	907	726	173	96
150	2500	2170	1085	870	208	115
200	3340	2900	1450	1160	278	153
250	4200	3650	1825	1460	349	193
300	5050	4400	2200	1760	420	232
350	5860	5100	2550	2040	488	270
400	6670	5800	2900	2320	555	306
450	7470	6500	3250	2600	620	344
500	8340	7250	3625	2900	693	383

APPENDIX E

List of Principal Symbols

Chapter 1

A = Area, meter^2 (m^2)
B = Magnetic flux density, webers/m^2 or tesla (Wb/m^2 or T)
B_g = Air-gap flux density, Wb/m^2 or T
B_m = Maximum flux density, Wb/m^2 or T
B_r = Residual flux density, Wb/m^2 or T
B_s = Saturation flux density, Wb/m^2 or T
e = Induced emf, volts (V)
F = Force, newtons (N)
$\mathcal{F}$ = Magnetomotive force or mmf, ampere-turns (At)
$\mathcal{F}_g$ = Air-gap mmf, At
f = Frequency, hertz (Hz)
H = Magnetic field intensity, At/m
H_c = Coercive force, At/m
H_g = Air-gap magnetic field intensity, At/m
I = Current, amperes (A)
k = Coupling coefficient
L = Inductance, henries (H)
L_l = Leakage inductance, H

337

l = Length, meters (m)
l_g = Air-gap length, m
M = Mutual inductance, H
N = Number of turns
P_e = Eddy-current loss density, watts/kilogram (W/kg)
P_h = Hysteresis loss density, W/kg
$\mathcal{P}$ = Permeance, H
R = Resistance, ohms (Ω)
$\mathcal{R}$ = Reluctance, H^{-1}
V = Dc (or steady ac) voltage, V
v = Instantaneous voltage, V
W_e = Electrical stored energy, watt-seconds or joules (W-s or J)
W_m = Magnetic stored energy, W-s or J
λ = Flux linkage, Wb
μ = Permeability, H/m
μ_0 = Permeability of free space, H/m
μ_d = Differential permeability, H/m
μ_i = Initial permeability, H/m
μ_r = Relative permeability
φ = Magnetic flux, Wb
φ_c = Core flux, Wb
φ_g = Air-gap flux, Wb
φ_l = Leakage flux, Wb
φ_m = Mutual flux, Wb

Chapter 2

Notes:

1. All the symbols pertinent to the exact equivalent circuit of the transformer are listed at the end of Section 2.6 and most of these symbols are not repeated here.

2. Subscripts 1 and 2 correspond to the primary and secondary of the transformer, respectively. The subscript e is used for equivalent values.

3. The superscripts prime (') and double-prime (") are used for equivalent values referred to the primary and the secondary of the transformer, respectively.

a = Turns ratio
E = Rms induced voltage, V
e = Instantaneous induced voltage, V
$\mathcal{F}$ = Magnetomotive force, At
f = Frequency, Hz
I = Current, A
N = Number of turns

P_c = Core loss, W; also used for conductively supplied power in an autotransformer
$\mathcal{P}$ = Permeance, H
R = Resistance, Ω
$\mathcal{R}$ = Reluctance, H^{-1}
t = Time, seconds (s)
V = Voltage, V
V_0 = No-load terminal voltage, V
V_t = Terminal voltage, V
X = Reactance, Ω
X_l = Leakage reactance, Ω
X_m = Magnetizing reactance, Ω
Z = Impedance, Ω
φ = Magnetic flux, Wb
φ_c = Mutual flux, Wb
φ_l = Leakage flux, Wb
φ_m = Maximum flux, Wb
ω = Angular frequency, rad/s

Chapter 3

a = Number of parallel paths
B = Flux density, Wb/m^2 or T
E = Induced emf in the armature, V
e = Instantaneous induced emf, V
I_a = Armature current, A
I_f = Shunt field current, A
I_t = Total current, A
i = Instantaneous current, A
k = A proportionality constant
$k_1 = Zp/60$ = a design constant
l = Length, m
N = Number of turns
n = Speed of rotation, revolutions/min (rpm)
p = Number of poles
R_a = Armature resistance, Ω
R_f = Shunt-field resistance, Ω
R_{se} = Series-field resistance, Ω
r = Radius, m
T_e = Developed torque, newton-meters (N-m)
u = Linear velocity, m/s
V_0 = No-load terminal voltage, V
V_f = Voltage across the shunt field winding, V

V_t = Terminal voltage, V
v = Instantaneous terminal voltage, V
Z = Number of armature conductors
φ = Flux per pole, Wb
ω_m = Angular velocity of rotation, rad/s

Chapter 4

E = Induced voltage (or no-load voltage), V
e = Instantaneous induced voltage, V
$\mathcal{F}$ = Magnetomotive force, At
$\hat{\mathcal{F}}$ = Maximum mmf, At
f = Frequency, Hz
I_a = Armature current, A
I_c = Circulating current, A
i = Instantaneous current, A
k_d = Distribution factor
k_p = Pitch factor
$k_w = k_p k_d$ = Winding factor
m = Number of phases
N = Number of turns
n_s = Synchronous speed, rpm
P = Number of poles
P_d = Power developed, W
p = Number of pole pairs
Q = Number of slots
q = Number of slots per pole per phase
R_a = Armature resistance, Ω
V_0 = No-load terminal voltage, V
V_t = Terminal voltage, V
X_d = Direct-axis synchronous reactance, Ω
X_q = Quadrature-axis synchronous reactance, Ω
X_s = Synchronous reactance, Ω
x'_d = Direct-axis transient reactance, Ω
x''_d = Quadrature-axis transient reactance, Ω
Z_s = Synchronous impedance, Ω
α = Slot angle, electrical degree
β = Coil span
δ = Power angle, electrical degree
θ = Electrical degree
θ_m = Mechanical degree
$\dot{\theta} = d\theta/dt$ = time rate of change of θ
τ = Pole pitch

τ'_d = Transient time constant, s
τ''_d = Subtransient time constant, s
φ = Flux per pole, Wb
  (*Note:* φ has also been used to designate the phase angle)
ω = Angular frequency, rad/s
ω_m = Mechanical angular velocity, rad/s

Chapter 5

E_2 = Rotor-induced voltage, V
e_2 = Instantaneous voltage induced in the rotor, V
f = Supply (or stator) frequency, Hz
f_r = Frequency of rotor voltage and current, Hz
I_1 = Stator current, A
I_2 = Rotor current, A
I'_2 = Rotor current referred to the stator, A
N = Number of turns
n = Rotor speed, rpm
n_r = Rotor synchronous speed at rotor frequency, rpm
n_s = Synchronous speed of the stator rotating field, rpm
P_d = Developed power, W
P_g = Power crossing the air gap, W
P_i = Input power, W
P_o = Output power, W
P_r = Rotational power loss, W
p = Number of poles
R_1 = Stator resistance, Ω
R_2 = Rotor resistance, Ω
R'_2 = Rotor resistance referred to the stator, Ω
R_m = Resistance designating the iron losses, Ω
s = Slip
T_e = Electromagnetic (or developed) torque, N-m
T_{ef} = Developed full-load torque, N-m
T_{es} = Developed starting torque, N-m
T_i = Torque developed by the inner cage, N-m
T_o = Torque developed by the outer cage, N-m
X_1 = Stator leakage reactance, Ω
X_2 = Rotor leakage reactance at standstill, Ω
X'_2 = Rotor leakage reactance referred to the stator, Ω
X_c = Capacitive reactance, Ω
X_m = Magnetizing reactance, Ω
Z_i = Impedance of the inner cage, Ω
Z_o = Impedance of the outer cage, Ω

φ_m = Maximum air-gap flux, Wb
ω = Stator input frequency, rad/s
ω_m = Rotor angular velocity, rad/s

Chapter 6

B = Flux density, Wb/m² or T
f or f_1 = Supply frequency, Hz
f_2 = Rotor frequency, Hz
I_1 = Stator input current, A
I_a = Armature current, A
I_m = Current in the main winding, A
I_r = Current in the reference winding, A
I_s = Current in the starting winding, A
L_d = Direct-axis inductance, H
L_q = Quadrature-axis inductance, H
n = Rotor speed, rpm
n_s = Synchronous speed, rpm
P_d = Developed power, W
p = Number of poles
R_1 = Stator resistance, Ω
R_2 = Rotor resistance, Ω
R_2' = Rotor resistance referred to the stator, Ω
s = Slip
s_b = Slip referred to the backward field
s_f = Slip referred to the forward field
T_e = Developed torque, N-m
t = Time, s
V_1 = Stator voltage, V
V_c = Voltage across the control winding, V
X_1 = Stator leakage reactance, Ω
X_2 = Rotor leakage reactance, Ω
X_2' = Rotor leakage referred to the stator, Ω
X_m = Magnetizing reactance, Ω
Z_b = Impedance of backward-field equivalent circuit, Ω
Z_f = Impedance of forward-field equivalent circuit, Ω
α = Phase angle between I_m and I_s
δ = Power angle
φ = Flux per pole, Wb
φ_m = Maximum flux, Wb
ω = Stator supply frequency, rad/s
ω_m = Rotor angular velocity, rad/s

Chapter 7

- a = Number of parallel paths
- C = Capacitance, farads (F)
- E = Induced voltage, V
- f = Supply frequency, Hz
- I_a = Armature current, A
- $k_1 = Zp/2\pi a$ = a design constant
- L = Inductance, H
- n = Speed, rpm
- p = Number of poles
- R_a = Armature resistance, Ω
- T = Chopping period, s
- T_e = Developed torque, N-m
- t_{off} = Off-time of a chopper, s
- t_{on} = On-time of a chopper, s
- V_o = Output voltage, V
- Z = Number of armature conductors
- α = Duty cycle
- φ = Flux per pole, Wb
- φ_1 = Phase angle between supply voltage and the fundamental component of supply current
- ω_m = Motor angular velocity, rad/s

Chapter 8

- I_{av} = Average current, A
- P_e = Electrical power, W
- P_m = Mechanical power, W
- p = Instantaneous power, W
- T = Temperature, °C
- T_{av} = Average torque, N-m
- T_m = Final temperature rise, °C
- t = Time, s
- V_{av} = Average voltage, V
- W = Energy, W-s or J
- η = Efficiency
- θ = Power factor angle
- τ = Thermal time constant, s
- Ω_{av} = Average angular velocity, rad/s

APPENDIX F

Answers to Selected Problems

CHAPTER 1

1.1 (a) 0.012 Wb; (b) zero; (c) 8.48 mWb. **1.2** 1.95 V. **1.3** (a) 800 At; (b) 2546.5 At/m. **1.4** (a) 0.32 T; (b) 0.512 mWb; (c) 1.562×10^6 H^{-1}; (d) 0.64×10^{-6} H. **1.6** (a) 1.257×10^{-4} H/m; (b) 1.18×10^{-4} H/m; (c) 1.08×10^{-4} H/m. **1.7** 21.89A. **1.8** 819 turns. **1.9** $B_{g1} = 0.1$T; $B_{g2} = 0.2$T; $B_{g3} = 0.05$T. **1.10** $B_{g3} = 0.1$T **1.11** $B_{g1} = 0$; $B_{g2} = 0.3$T. **1.12** $L_{11} = 1.34$ mH; $L_{22} = 0.75$ mH; $L_{33} = 3$ mH; $L_{12} = 0.5$ mH; $L_{13} = 1.0$ mH. **1.13** 115.2 W. **1.14** 0.875. **1.15** 1 T. **1.16** 3.2 mH. **1.17** 12.57 mH. **1.18** 3.81 A. **1.19** 0.533 J. **1.20** 0.229 J. **1.21** (a) 214.2 W; (b) 133.3 W; (c) 178.5 W. **1.22** 0.77 T.

CHAPTER 2

2.2 (a) 2; (b) (i) 22.73 A, 45.45 A; (ii) 10.91 A, 21.82 A. **2.3** (a) 5.5 A; (b) 40 Ω; (c) 1210 W. **2.4** 16.5 mWb. **2.5** $X_m = 187.6$ Ω; $R_c = 202.1$ Ω. **2.7** 98.4%. **2.11** (a) 97.4%; (b) 96.8%. **2.12** 96.4%. **2.13** (a) 98.39%; (b) 98.12%. **2.14** $X_m = 891.15$ Ω; $R'_e = 0.0794$ Ω; $X'_e = 0.0398$ Ω. **2.15** 450.9 V; 97.58%. **2.16** 94.31%. **2.17** 65.45 A. **2.18** 98.09%; 23.17 A. **2.19** 57.47 kW; 38.53 kW. **2.20** 1.67 A; 29.17 A; 1.14 A; 20 A. **2.21** $I_{1\text{ line}} = 3.15$ A; $I_{2\text{ line}} = 17.32$ A. **2.22** 96.36%. **2.23** 86.6 kVA. **2.25** $I_{2\text{ line}} = 49.21$ A. **2.27** 48 kW.

345

CHAPTER 3

3.3 (a) 36 V; (b) 67.5 V. **3.4** 0.03 Wb. **3.5** 114.6 N-m. **3.6** 250.24 V. **3.7** (a) 96,875 W; (b) 98,009 W. **3.8** 396.6 rpm. **3.9** 0.1215 Ω. **3.11** 11,428 W. **3.12** 1200 W. **3.15** (a) 254 V; (b) 226 V. **3.16** (a) 46.89 kW. **3.17** 90.6%. **3.18** (a) 188.6 N-m; (b) 93.17%. **3.19** 802.8 rpm; **3.20** 1035.8 rpm. **3.21** 692.8 rpm. **3.22** (a) 1129 rpm; (b) 108.9 N-m; (c) 88.3%. **3.23** 1297.6 rpm. **3.24** 1020 rpm; 84.3%. **3.25** 226.52 Ω. **3.26** 514 rpm. **3.27** (a) 0.704 Ω; (b) 5117.5 N-m. **3.28** 753.5 rpm.

CHAPTER 4

4.1 1000 rpm. **4.2** 0.966; 0.966. **4.3** 1841.4 V. **4.4** 15 mWb. **4.6** 4583.3 V. **4.7** 50.72 percent. **4.8** 7.44°; -8.16 percent. **4.9** 164.7 percent. **4.10** 11.54 kV. **4.11** (a) 22 MW; (b) 1566.4 A; (c) 0.737. **4.12** 93.2 percent. **4.13** 260.5 V/phase. **4.14** (a) 10.48 kV; (b) 11.04 kV; (c) 11.56 kV. **4.15** $\cos \varphi = 1$. **4.16** $\cos \varphi = 0.53$; 196.9A. **4.17** (a) 786.4 kW; (b) 6260 N-m; (c) 40.5°. **4.18** 107.67 kVA. **4.19** 150.64 kVa.

CHAPTER 5

5.1 4 percent. **5.2** 25 percent. **5.3** 144.35 N-m. **5.4** 0.4 Ω. **5.5** 960 rpm. **5.6** 3341 rpm. **5.7** 500 W. **5.8** 76.95 percent. **5.9** 3.27 percent. **5.10** (a) 24,084 W; (b) 187.5 N-m; (c) 191.7 N-m. **5.11** 3.2 Ω. **5.12** 20,888 W. **5.13** (a) 14.61A; (b) 0.84. **5.14** 547.14 Ω; 56.4 Ω. **5.15** $R_2 = 0.433 \ \Omega$; $X_1 = X_2 = 1$. **5.17** (a) 298.14 V; (b) 111.8 A. **5.18** 41.67 percent of full-load torque. **5.19** (a) 4.69 percent; (b) $V_{st} = 0.5V_f$. **5.20** 32.

CHAPTER 6

6.1 16.67 percent; 183.3 percent. **6.2** 2.28 A; 78.32 W; 0.219 N-m. **6.4** 47.8 percent. **6.5** 6.92. **6.6** $X_1 = 9.57 \ \Omega$; $X_m = 13.68 \ \Omega$. **6.7** 0.021 N-m. **6.9** 12,000 rpm. **6.10** 0.794 hp; 1.57 N-m. **6.11** 0.38 N-m. **6.12** 88.37 percent.

CHAPTER 7

7.1 43.2 V. **7.2** 0.286. **7.5** 18.52 A. **7.6** 1000 pulses/sec. **7.7** 290.9 μF; 11.8 mH. **7.8** 239 rpm. **7.9** 576 rpm; 80 N-m.

CHAPTER 8

8.1 89.2 percent. **8.2** 40.8 W. **8.3** 94 percent; 94 percent. **8.4** 0.6 year. **8.5** 49.6°C.

INDEX

A

AC armature windings
  chorded, 161
  distribution factor, 169
  double layer, 159
  emf equation, 109, 172
  fractional pitch, 161
  pitch factor, 170
  winding factor, 170
AC commutator motors, 249–252
AC motors, controllers, 283–292
Adjustable speed drive, 264
Ampere's law, 3
Angle, electrical, 165
Armature reaction, 126
Autotransformer, 48
AWG, 326

B

Back emf, 124
BH characteristic, 11, 13, 15
Bll rule, 3

C

Choppers, 273–278
Coercive force, 14
Commutation, line, 279
Commutator, 110, 113
Contactors, 139
Control for efficiency, 263
Controller induced losses, 292
Converters, 279–283
Core, 15
Core loss, 26
Coupling coefficient, 37
Current transformer, 93

D

Darlington, 266
DC machines, 103
  airgap fields, 127
  applications, 147
  armature reaction, 126
  armature windings, 111

DC machines (cont.)
   automatic starters, 139
   back emf, 124
   characteristics, 133, 136
   classification, 119, 120
   commutation, 109, 129
   commutator, 109–110, 129
   compensating winding, 128
   construction, 111
   control, 137
   conventional, 107
   critical field resistance, 132
   description, 111–118
   efficiency, 142
   emf equation, 109, 121
   energy conversion, 103
   equalizer connections, 149
   field poles, 111, 117
   field-resistance line, 132
   field winding, 111
   generator, 103
   geometric neutral plane, 112, 126
   governing equations, 121–123
   heteropolar, 107
   interpole, 130
   lap winding, 112, 113
   load sharing, 148
   losses, 142
   magnetization curve, 132, 134
   manual starters, 138
   motor, 103
   motor controllers, 270
   parallel operation, 147
   power flow, 143
   reactance voltage, 130
   separately excited, 120, 124, 134, 135
   series motors, 120, 137
   shunt machine, 120, 131, 135, 137
   slots, 112, 113
   speed equation, 124, 125
   starting, 137
   teeth, 112, 113
   tests, 144
   torque, 122
   voltage build-up, 131
   Ward-Leonard system, 142
   wave winding, 112, 113
Diamagnetic material, 10
Direct-axis inductance, 245
Duty cycle, 302–306

E

Eddy current, 26
   loss, 26
Electric machines
   economics, 308
   insulation, 317
   motor selection, 313
   name plate ratings, 299
   NEMA classification, 306
   ratings, 306
   thermal considerations, 310
Electrical angle, 165
emf
   equation, 107, 172
   induced, 24, 25
Energy conversion, 109
Energy stored in magnetic field, 38
Equivalent circuits, 30, 61, 68
Exciting current, 62

F

Faraday's disc, 104
Faraday's law, 24, 25, 104
Ferrimagnetic materials, 10
Ferromagnetic materials, 10
Field pole, 112, 117
Flux cutting rule, 105
Flux linkage, 25
Flux, magnetic, 3
Fractional-pitch winding, 170
Frame size, 306, 307
Free wheeling diode, 280
Fringing, 24
Full-pitch winding, 170

G

GTO, 266

H

Homopolar machines, 104
Hysteresis loop, 14
Hysteresis loss, 15, 26
Hysteresis machines, 246

I

Impedance matching, 47
Inductance
   definition, 32

leakage, 32
mutual, 32, 36
self, 32
Induction machines, 199
Induction motor, 202
   action, 202
   blocked rotor test, 213
   brushless, 199
   cage type, 199
   capacitor-start, 243
   characteristics, 207–212
   double cage, 222–224
   efficiency, 208, 211
   energy efficient motors, 228
   equivalent circuit, 206, 207, 208, 214
   generator, 225–228
   leakage reactance, 206
   magnetizing reactance, 206
   no-load test, 212
   performance calculations, 207
   performance criteria, 215
   power across air-gap, 208
   power developed, 208
   power flow, 208
   rotating magnetic field, 166
   rotor equivalent circuit, 206
   rotor resistance, 206
   shaded pole, 243
   single phase, 235–243
   slip, 203
   slip frequency, 203
   slip ring, 199
   speed control, 217–219
   split phase, 242
   starting, 220–224
   synchronous speed, 168, 203
   tests, 212
   torque, 209
   wound rotor, 199
Insulation, 317
Inverters, 294–292

L

Laminations, 27
Lap winding, 114, 115
Leakage flux, 33
Leakage inductance, 33
Leakage reactance, 38
Left-hand rule, 3
Lenz's law, 25

M

Machine ratings, 299, 306
Magnet, permanent, 1
Magnet wire, 326
Magnetic circuits, 1, 6, 15, 23
Magnetic field, 2
   energy in, 38
Magnetic field intensity, 6
Magnetic flux, 3
   sources of, 5
Magnetic material, 9
   diamagnetic, 10
   ferrimagnetic, 10
   ferromagnetic, 10
   hard, 10
   paramagnetic, 10
   soft, 10
Magnetic saturation, 11
Magnetic stored energy, 38
Magnetization (or BH) characteristics, 11, 13, 15
Magnetizing reactance, 38
Magnetomotive force, 5
Mean length, 19
mmf definition, 5
Motor starting, 137, 177, 220

N

Nameplate rating, 299
NEMA (National Electrical Manufacturers Association), 306–308

O

Ohm's law for magnetic circuits, 17
Oil, insulating, 49

P

Paramagnetic material, 10
Permeability, 9
   differential, 11
   free space, 9, 10
   incremental, 11
   initial, 10
   relative, 10
   relative amplitude, 10
Permeance, definition, 17

Phase angle error, 95
Phase sequence, 81, 185
Pole, magnetic, 2
Pole pitch, 112
Power semiconductors, 265–270
Power transistor, 268
Pulse width modulation (PWM), 286

Q

Quadrature-axis inductance, 245

R

Rating, machine, 299, 306
Reluctance, definition, 17
Reluctance machines, 244
Reluctance torque, 244
Repulsion motors, 252
Residual flux density, 133
Residual magnetism, 133
Right-hand rule, 3, 5, 106
Rotating magnetic field, 165–167
Rotating rectifier machine, 158

S

Saturation, 11
Saturation flux density, 14
Silicon controlled rectifier, 267
Silicon rectifier, 265
Slip, 203
Solid-state control, 270–289
Stacking factor, 27, 28
Stepper motor, 255
Symbols and units, 327, 337–343
Synchronous impedance, 173, 190
Synchronous machines
   armature mmf, 163
   armature resistance, 173
   armature windings, 159, 160, 161
   compounding curves, 179
   construction, 156
   cooling, 162
   damper bars, 162
   direct-axis reactance, 182
   equivalent circuits, 176
   exciter, 156
   generator operation, 171–175
   hysteresis motor, 246
   motor operation, 176, 178
   overexcited, 178
   parallel operation, 184
   phasing out, 186
   pilot exciter, 156
   power angle, 175, 182, 183
   quadrature-axis reactance, 182
   reactances, 173, 193
   reluctance motor, 244
   reluctance torque, 177
   rotating rectifier, 156, 158
   round rotor, 157, 171
   salient pole, 159, 181
   starting, 177
   subtransient reactance, 191
   sudden-short, 191
   synchronizing, 185
   synchronous impedance, 173, 190
   synchronous reactance, 173, 190
   synchronous speed, 168
   tests, 190–192
   transient, 191–192
   transient reactance, 191
   two-reaction theory, 182
   underexcited, 178
   V curves, 179
   voltage regulation, 174
Synchronous reactance, 173, 190
Synchronous speed, 167
Synchroscope, 186

T

Three-phase transformers, 80
Thyristor, 267
Toroid, 44
Torque angle, 175
Transformer, 47
   audio, 53
   auto, 48, 53, 88
   classification, 52–53
   connections, 80–86
   construction, 48
   core, 49
   core loss, 39, 62, 71
   core loss resistance, 62, 71
   current, 93
   delta-wye, 84
   efficiency, 65
   electronics, 52
   emf equation, 55

equivalent circuit, 62, 63, 68, 94
equivalent impedance, 62, 63, 68
excitation characteristic, 95
harmonics, 95
ideal, 53–60
impedance, 47
impedance matching, 47
instrument, 52, 93
leakage reactance, 62
magnetizing current, 62
magnetizing reactance, 62
non-ideal, 61
oil, 49
open-circuit test, 71
open delta, 85
parallel operation of, 77, 87
polarity, 75
polarity test, 75
potential, 93
pulse, 53
saturation, 95
Scott connection, 87
short-circuit test, 72
specialty, 53
tests, 71, 72
three-phase, 80
turns ratio, 54
voltage regulation, 65
winding resistance, 61
windings, 48
Transistor, 270
Two-phase motors, 253–255

U

Unit conversion, 327
Universal motor, 249

W

Wave winding, 114, 116
Windings
  dc, 114–116
  simplex, 114–116
  transformer, 48
Wire table, 326